Innovation in Weapon Systems

What Can History Teach Us?

Leonard Gaston

This book is dedicated to my father. A strong and noble man, dedicated to the well being of his loyal wife and his children. A brilliant and talented individual, he was constrained by circumstances to spend his working life in hard farm labor. He raised my sisters and two older brothers to maturity on a dust bowl farm in Kansas. Those young men left the farm in the opening years of World War II, earned silver wings and commissions in the Army Air Corps, and completed combat tours in the skies over Europe. He saw his two decorated sons return safely before his untimely death in a railroad accident.

Personal acknowledgments

Dr. Greg, Dr. Bob, and Renee'
Thank you for your never ending help and encouragement.
Janet, the best daughter in law a man could have,
thank you for your patience and concern.

I am grateful to my late wife Betty who spent long hours at her typewriter, turning hand written drafts into reports, papers, thesis, dissertation, and, of particular relevance here, an organized record of my first attempts at digging into history. Those pages furnished the springboard and inspiration for this book.

....

Professional Acknowledgments

I am indebted to Brenda Van Niekerk whose incredible skills in formatting and computer wizardry continue to amaze me and are equaled by her patience and understanding. Brenda has proven to be a wonderful person to work with and can be contacted at brendavniekerk@hotmail.com

I am likewise grateful to Laura Shinn whose skills in graphic arts have graced my books. I am in awe of Laura's talent. She can be reached at
http://laurashinn.yolasite.com

Table of Contents

CHAPTER 1 1

    THE CONTINUED EXISTENCE OF WAR 1

CHAPTER 2 4

    TECHNOLOGY MAKES USEFUL WEAPONS: APPROPRIATE DOCTRINE MAKES THOSE WEAPONS USEFUL 4

        Prediction of Availability, Cost, and Effectiveness 4

        The Importance of Range of Action 5

        The "master weapon" changed strategic thought. 7

        Fuller's Five Attributes Sorted out: Range of Action, an Effective Shield, and Portability 7

CHAPTER 3 11

    WARFARE: SOMETIMES LIMITED, SOMETIMES UNLIMITED. 11

        Unlimited War in the Ancient World 11

        Greek Limited War 11

        Conditions Tending to Limit War in the Middle Ages 12

        The Impact of Mass Armies 13

        Modern War 14

CHAPTER 4 16

    THE SEARCH FOR RANGE: EARLY WEAPONS 16

        Early Use of the Bow 16

        Byzantine Use of the Bow and Other Weapons 19

        The Eastern Roman View of War 20

            The Byzantine Horse-Archer 21

        Byzantine Naval Superiority 22

The Short Bow in Europe 24
The Battle of Hastings 25
The Crossbow 26
The Longbow In the Hundred Year's War 26
Crecy, Potiers, and Agincourt 27
Eventual English Defeat In France 29

CHAPTER 5 30
THE SEARCH FOR RANGE: THE AGE OF GUNPOWDER 30
Catapults and Gunpowder 30
Early Cannon 31
Improvements to the First Cannon 32
Cast Iron 32
Light Weight Cannon 32
Ballistics Developments 32
The Siege of Constantinople 33
French Cannon in Italy – The Doom of Fortifications? 34
Artillery In the Open Field – Temporarily a Master Weapon 35
Uneven Progress In Artillery 36
Breech-Loading Cannon and The Franco-Prussian War 37
The British Establishment Clings to Muzzle-Loading Artillery 37
Control of Artillery Recoil 38
Delays In Use of New Technology 39
Gun Powder in Infantry Weapons 39
The Spanish Matchlock 39
A Replacement for the Matchlock: The Wheel Lock 40
The Flintlock 41
Percussion Ignition 41
The Prussian Needle Gun 42
The Minie Ball 43
The Development of Advanced Small Arms 44
Early Concepts in Repeating Firearms 44

  James Puckle's 1718 "Machine Gun" 45

  Samuel Colt's Revolver 45

  The Repeating Rifle 46

  The Ultimate Destroyer: Maxim's Machine Gun 46

  A Near-Master Weapon: Its Impact 47

  French Obsession with the Offensive 48

CHAPTER 6 50

THE SEARCH FOR A BETTER SHIELD 50

Metal Armor and Arms 50

  Primitive Metallurgy 50

  Steel 51

  Shield and Body Armor 52

  Armor in the Middle Ages 53

  The Developing Technology of Medieval Body Armor 53

Fortifications 55

  The Viking Threat to Europe 55

  Development of the Castle 56

  Inherent Strength In a Strong Defensive Posture 56

Spear As Shield 57

  Stakes and Spears 58

  Limitation of Spear as Shield 60

  The Importance of the Shield 60

The Maginot Line 60

CHAPTER 7 62

THE QUEST FOR MOBILITY 62

Early Attempts to Achieve Mobility 62

The Horse 64

  The Horse As Cavalry Mount 64

  The Horse as a Master Weapon 65

Ships Powered by Oar and Sail 66

CHAPTER 8 69

SAIL AND GUNPOWDER 69

The Spanish Monarch Determines to Invade England 69
Royal Rivalries 70
The Armada 71
The English Fleet 72
Innovative Ship Design 73
England's Lack of Preparedness 74
Indecision 75
The Approach of the Armada 77
Battle 78
The Price of Victory 81

CHAPTER 9 82
STEAM, STEEL, AND MODERN WAR 82
Early Opposition to Steam Power for Ships 83
The Triumph of Steam 84
The Steam Engine on Land 84
The Internal Combustion Engine on Land 85
Gasoline Engine plus Chain Track 85
Bureaucratic Indifference 86
A New Weapon Poorly Used 86
Tank Planning Between Wars 87
The Introduction of Airborne Weapons 88
The Steel Wedges of Monseigneur Garros 88
The Fokker: Scourge of the Skies 88
Early developments in rocket propulsion 90

CHAPTER 10 92
CHALLENGE NUMBER ONE – OVERCOMING THE POWER OF PREVAILING FASHION 92
Some Examples of the Power of Fashion 93
The importance of tradition 93
Drop Tanks for Escort Fighters in World War II 94
The Idea of Using Drop Tanks Was Not New 96

It Can Be Difficult to Think About New Things. 97

Fashion versus technology in the Modern Era 98

The Need for Continued Innovation 100

CHAPTER 11 102

CHALLENGE NUBER TWO: LOOKING ACCURATELY INTO THE FUTURE 102

Dominance Is Important, But Always Temporary 102

The present importance of technological advantage 103

Five Common fallacies about technology 104

The difficulty of making accurate predictions 105

Technology Forecasting 107

First – Supporting technologies. 107

Second – Identification of promising areas of technology: The technological growth curve, or S – curve. 108

Third – The shift from optimism to pessimism. 112

Fourth – Is invention inevitable? 113

Invention A can eliminate the need for Invention B. 116

CHAPTER 12 117

TODAY'S REALITY AND TODAY'S FASHION 117

Trends in U.S. Defensive Forces Versus Potential Rivals 117

The U.S. Depends on Deterrence. 118

The U.S. Cold War Buildup and Its Demise. 118

New Russian Missile Submarine 120

Is Our Concept of Deterrence Shared by the Russians? 120

Nuclear Treaties between the U.S. and the Soviet Union/Russia 122

Mobile Rail Systems 124

U.S. Strategic Forces Have Been Reduced in Numbers 125

Strategic Forces in 1985 125

U.S. Intercontinental Ballistic Missiles in 1985 125

Soviet Intercontinental Ballistic Missiles in 1985 126

U.S. SLBMs in 1985 126

Soviet SLBMs in 1985 127

Current U.S. Strategic Forces 127

Current Russian Strategic Forces [21] 128

Land-based 128

Sea-based 128

Air-based 128

Hot Launch versus Cold Launch 129

Trends In Modernization 130

Warhead Modernization 130

China 130

Russian Modernization and New START 131

More Modern Launch Vehicles 131

Other Countries 133

Missile Defense Systems 134

Unmanned Aerial Vehicles 135

Satellites, data links, and anti satellite weapons 135

Summary 136

CHAPTER 13 138

THE IMPACT OF THE NUCLEAR BOMB 138

The Battle That Never Was 138

The Purple Heart 138

U.S. Leaders Planned to Invade Japan to End the War 139

The Navy Insisted Invasion Was Not Required 139

The U.S. Invasion Plan 141

Japanese Plans 141

Kamikaze Attacks 142

For Troops That Made It Ashore The Beaches Would Be Deathtraps 143

Japanese Civilians Would Fight to the Death. 144

What Did Use of This Weapon Achieve? 144

An Ironic Historical Note 146

    Superior Technology Has Brought the U.S. Safety, What Might Inferior Technology Bring? 146

        The Trap of Being Number One 147

APPENDIX 148

A WAR AGAINST TERROR 148

The Longest Conflict in Human History 149

    Imperialistic and Colonialist Expansion of the Muslim Empire 150

Current Trends in Terrorism 152

Terrorist Groups 153

    The Islamic State of Iraq and Syria (ISIS) 153

    Did U.S. military action open the way for expansion of ISIS into Libya? 155

    Boko Haram 157

    Islamic Revolutionary Guard Corps – Quds Force 157

    Haqqani Network 158

    Kataib Hezbollah 158

    Hezbollah 159

    The Muslim Brotherhood 159

    Hamas 160

    Al-Qaeda 161

    Other Organizations 162

The Threat to the United States 162

Terrorists Have Some Significant Advantages 163

    Use of Modern Technology 163

    Use of The Internet to Recruit Domestic Terrorists 165

    More Radicalized Muslims Means More Terrorist Attacks 167

How Do We Defend Ourselves? 168

Reference Notes 171

    Chapter 1 The Continued Existence of War 171

    Chapter 2 Technology Makes Useful Weapons, Appropriate Doctrine Makes Those Weapons Useful 171

    Chapter 3 Warfare: Sometimes Limited, Sometimes Unlimited 171

    Chapter 4 The Search for Range: Early Weapons 172

- Chapter 5 The Search for Range The Age of Gunpowder 173
- Chapter 6 The Search for a Better Shield 175
- Chapter 7 The Quest for Mobility 176
- Chapter 8 Sail and Gunpowder 176
- Chapter 9 Steam, Steel, and Modern War 177
- Chapter 10 Challenge Number One – Overcoming the Power of Prevailing Fashion 178
- Chapter 11 Challenge Number Two – Looking Accurately into the Future 179
- Chapter 12 Today's Reality and Today's Fashion 181
- Chapter 13 The Impact of the Nuclear Bomb 185

Notes for Appendix The War Against Terror 187

"Even the most cursory survey of military history substantiates the premise that superior weapons give their users an advantage favoring victory. A somewhat closer study of military history shows that new and more effective weapons have generally been adopted only slowly in spite of their obvious advantages. Since the character of contemporary weapons is such that their production as their use can dislocate whole economies, it is probably not too much to suggest that the survival of entire cultures may hinge upon an ability to perfect superior weapons and exploit them fully. Survival itself, then, appears to depend on speed in both the development and the utilization of weapons."

> I. B. Holley, Jr.

My first wish is to see this plague of mankind, war, banished from the earth.

> George Washington

"There is nothing so likely to produce peace as to be well prepared to meet the enemy"

> George Washington

The men of Normandy had faith that what they were doing was right... There is a profound, moral difference between the use of force for liberation and the use of force for conquest. You were here to liberate, not to conquer.

> President Ronald Reagan, June 6, 1984,
> At the U.S. Ranger Monument at Pointe
> du Hoc, France, On the 40th Anniversary
> of D-Day

CHAPTER 1

THE CONTINUED EXISTENCE OF WAR

War is about killing people and destroying things. In the beginning a group of men might say, "We are better" (stronger, smarter, swifter, or more favored by the gods) "than they are." "Let us go over there and kill them, take their stuff, and take their women and children as slaves." On the other side, threatened with aggression, men might say "We must fight to defend ourselves. Otherwise they will kill us all, take our stuff, and make our wives and children slaves."

In today's world, conquest of a nation by an aggressor might or might not result in massacre, depending on who the aggressor was. It would certainly result in loss of all freedoms and severe degradation of manner of living. In any case the potential victim faces the stark choice: submit or fight a defensive war.

Social organizations develop and leaders rise up. A leader may desire fame, territory, wealth, power, or something else – but when a Cyrus, a Darius, an Alexander, a Caesar, or a Hitler can move a people – through force, charisma, or ideas – to fight for the objective the leader desires, a war of aggression ensues. If a society is to survive, his potential victims have no choice but to take up arms to defend themselves. The tiny state of Israel, threatened by invasion from hostile Arab powers since its birth in 1948, furnishes a powerful present-day example.

Whatever the reason for combat, history shows us that technological innovation can have a significant impact on the outcome. It can lead to quick victory or shield combatants from harm. Superior technology can prevent aggression by making its cost prohibitive.

Since World War II the United States has managed to convince 'the other guy' that if he tried to destroy us, we would destroy him. U.S. possession of thermonuclear weapons and delivery vehicles has allowed it to rely on the doctrine of Mutual Assured Destruction to prevent general war. Although no immutable law of nature assures us it will remain valid forever, Mutual Assured Destruction (MAD) can claim past effectiveness against a large nation-state with a lot to lose. But we have moved from a bipolar world to a world where many parties can acquire and employ weapons of mass destruction. A threat of massive retaliation has little or no value against smaller entities that believe they can escape identification as the source of an attack or non-state groups whose decision agents have unknown locations or who simply consider damage to us more important than their own lives.

In the years following World War II, when the U.S. relied on a Balance of Terror to prevent outbreak of World War III, hostile powers made effective use of asymmetrical warfare against the United States. In Vietnam for example, its opponents enticed the U.S. - a country well equipped to fight a capital intensive war - into a labor intensive war on the ground. The resulting toll of death and injury to our fighting men, was tragic (58,000 deaths and another 150,000 wounded), and - as it turned out - pointless. Today there is no South Vietnam. Our country carries on peaceful trade with Vietnam, a Communist entity.

History demonstrates that institutional factors can determine whether available technologies are promptly applied, delayed, or never used. It demonstrates also that whatever the technology available, clear thought, unhindered by the blinders of custom is necessary to develop the best use of that technology.

This book provides a review of selected historical incidents and a few concluding thoughts. The reader is left to answer the question: How can the U.S. use developing technology to better assure the safety of its people?

Many scholars have studied the history of warfare; their findings fill volumes. This book can provide only a few examples of technological innovation applied to warfare. Where it appeared necessary to convey the full sense of what cited scholars were saying, their words are presented verbatim,

The amount of historical material available is overwhelming. This little book simply discusses four things (1) Selected historical innovations that had significant impact on armed conflict, (2) institutional factors that facilitated or inhibited their adoption, (3) difficulties in understanding what technologies the future might bring and how they could be used, and (4) the threats that exist today.

As we look at innovation across the canvas of history we better understand the huge impact that innovative technology can have. We recognize that our country must carefully choose the best ways to use advancing technology to protect itself. We realize that America must husband its resources and use the most efficient means possible to protect our society – not only from hostile nation states – but from other entities that will destroy us if they can.

Every up-to-date dictionary should say that 'peace' and 'war' mean the same thing, now in posse, now in actu. It may even be said that the intensely sharp competitive preparation for war by the nation is the real war, permanent, unceasing, and that battles are only a sort of public verification of mastery gained during the 'peace' intervals.

William James

CHAPTER 2

TECHNOLOGY MAKES USEFUL WEAPONS: APPROPRIATE DOCTRINE MAKES THOSE WEAPONS USEFUL

Careful planning of future weapon systems, followed by adequate funding to develop and field them, can sometimes prevent actual conflict. If it cannot, sound planning should lay the groundwork for reduction of losses in the in actu phase described by James.

Prediction of Availability, Cost, and Effectiveness

Effective planning for future weapon systems must assess (1) The progress of technology, (2) Measures necessary to put technology into a working weapon system, and (3) The future effectiveness of that system. This job is difficult when it involves incremental advances in familiar technologies. It is more difficult when it requires forecasting how rapidly a new technology will develop. Judging future effectiveness may be most difficult of all, because the best ways to apply the new technology may not be obvious.

Since the development of weapons began, as new technologies have arrived on the scene, successful application of these technologies has changed the course of battles, wars, and empires. Lack of successful application has had a negative impact on those who overlooked or misapplied them. At times, new weapons have been deliberately sought and quickly applied, as in the case of British development of the exploding shell to fire at opposing wooden warships during the siege of Gibraltar. At other times new weapons have appeared on the scene only to be ignored until years later. When new weapons were available and an attempt made to use them, all too often the initial use was inappropriate and ineffective. In the words of a prominent historian, Major General D. K. Palit:

> *If there is one recurrent theme that emerges from an analysis of tactical evolution, it is the resistance of military conservatism to the adoption of new tactical doctrine at each state of the history of warfare. Whether it was the introduction of the longbow or the musket, the development of modern artillery or the invention of the tank, re-adjustments in tactical methods lagged far behind the impact of the weapons themselves. In some cases ... not only did conservative opinion resist the introduction of the new weapons, but even when they were finally taken into use, their full potential was not realized for a considerable period.*[1]

A complete history of the development of weapons would fill a library of books. What is attempted here is presentation of a few highlights, with brief supplemental discussion. Only three general chains of development are treated, and these with less than complete coverage. The first, and perhaps the most fundamental, is the search for increased range of effect, as typified by the short bow, the long bow, and the eventual uneven development of gunpowder weapons. The other two, complementary to the first, are the search for a better shield, and the quest for increased mobility.

The Importance of Range of Action

There is an old army saying: "If the enemy is in range, so are you."

J.C.F. Fuller has written that the effectiveness of any weapon depends on the presence or absence of five characteristics. He lists range of action as the most significant.

    1. Range of Action

Chapter 2 – Technology Makes Useful Weapons: Appropriate Doctrine Makes Those Weapons Useful

    2. Striking power
    3. Accuracy of aim
    4. Volume of fire
    5. Portability

He states that the first, range of action, is dominant. The weapon of superior reach or range should be central to tactics.

> *Thus, should a group of fighters be armed with bows, spears, and swords, it is around the arrow that tactics should be shaped; if with cannon, muskets, and pikes, then around the cannon, and if with aircraft, artillery, and rifles, then around the airplane. The dominant weapon is not necessarily the more powerful, the more accurate, the more blow-dealing, or the more portable; it is the weapon which, on account of its superior range, can be brought into action first, and under the protective cover of which all other weapons according to their respective powers and limitations can be brought into play.[1]*

The greater the striking power, accuracy, volume of fire, and portability of the weapon with extended range of action, the more dominant this weapon. Writing prior to the first use of atomic weapons, General Fuller even then described the "air-carried bomb", with its range of action, striking power, and portability, as the dominant weapon of the time. Its partially offsetting weaknesses, he felt, were a low accuracy of aim of the aircraft and a low volume of fire due to the aircraft's need, once its bombs had been "fired", to return to its base to reload. Prophetically, he said:

> *Could these limitations be surmounted, then the airplane would become what may be called "a master weapon" – that is, a weapon which monopolizes fighting power.[2]*

In the absence of defenses to fully exploit its relative vulnerability, the aircraft carrying nuclear bombs briefly did become that master weapon, to be supplanted, in turn, by the new master weapon, the thermonuclear tipped Intercontinental Ballistic Missile.

The "master weapon" changed strategic thought.

The existence of this new master weapon did, in fact, so dominate thinking that it led not only to official adoption of the doctrine of "Mutual Assured Destruction", but to a belief that defense was impossible, use of nuclear weapons was unthinkable, and a threat of massive retaliation therefore was a sufficient doctrine for national defense, now and forever.

Reality is not that simple. It appears this school of thought is not necessarily held in high esteem by leaders in Russia, China, North Korea, or Iran. (Note: this list of unbelievers is subject to expansion with the passage of time.)

In the opening years of the 21st Century U.S. forces have been stretched thin in deployments around the globe and asked 'to do more with less'. World-wide basing and continuing wars in the Middle East consume scarce resources. Potential enemies develop new strategic weapons. Terrorist groups openly tell us they will destroy us. Should we wonder if present policies may need modification and some new thinking might be in order? And that the long in the tooth doctrine of Mutual Assured Destruction is increasingly best described by its acronym (MAD)?

Fuller's Five Attributes Sorted out: Range of Action, an Effective Shield, and Portability

Range of action is obviously important. If individual, group, or country A can make a significant impact on individual, group, or country B – without B being able to reach back and impact A, then A can operate with impunity.

Striking power fundamentally refers to the amount of destructive energy that A can deposit on B in a short period of time, possibly before B can respond. (In the modern world it might apply also to disruption of computers, data systems, or other vital infrastructure without physical destruction.) Historically this could have been done by moving foot soldiers very rapidly to a place where they outnumbered the opponent's foot soldiers, and then pouring destructive energy on the force that has been outmaneuvered – whether by club, battle ax, arrow, or gunpowder launched projectile. A cavalry charge might be used for this purpose. If fortifications are the target than the striking power might be provided by artillery of some kind.

Chapter 2 – Technology Makes Useful Weapons: Appropriate Doctrine Makes Those Weapons Useful

Accuracy of aim simply means that if A is launching destructive energy in the direction of B it has no impact unless A can actually hit B with the energy. Historically, when individuals fought against individuals, accuracy of aim was improved by getting closer to the target, but that would make A vulnerable to retaliation by B – which made range of action very important.

Volume of fire could be defined as striking power by another name – the ability to severely impact the target in a very short period of time. Traditionally this was done by having more systems, which meant more humans, or later, more tanks, more artillery, or more airplanes, at a given location to generate and direct destructive energy at the target. This, we might say, could be a function of portability.

Portability is the ability to move something from place to place quickly and easily: typically some means A could use to move a source of destructive energy to a position where that energy could be deposited on B, or an ability by B to evade hostile action.

Range of Action, an Effective Shield, and Portability

Based on this background, each attribute discussed here has been placed into one of three categories: the search for range, the need for a better shield, and the quest for portability.

This chapter has briefly discussed critical weapon systems characteristics

Chapter 3 presents a historical overview of the subject of warfare. This overview of a small sample of the battles that have taken place through the ages makes us appreciate the sincere desire of many people – utopian though it may be – to eliminate from the face of the earth something as horrible as warfare, the deliberate attempt by one group to kill or maim the humans in another group. Nevertheless, if we hope to prevent war or alleviate its effects, it is necessary to better understand the impact of technology.

Chapters 4 and 5 are devoted to the search for Range of Action. What A would desire is the ability to deposit destructive energy on B without B being able, in return, to deposit destructive energy on A. This could be accomplished if A has a weapon with superior range. So range of action – as Fuller points out – is the dominant characteristic of a weapon. It will be discussed first. The greater the amount of destructive energy that can be deposited at long range, the greater the impact. The invention of gunpowder increased range of action and also increased the amount of destructive energy that could be projected (heavier projectiles, exploding projectiles).

Chapter 6 treats The Search for a Better Shield. Although the shield is not an offensive weapon, the search for ways to blunt an enemy's attack has been a historical imperative.

Chapter 7 discusses the search for mobility. The extent to which portability or mobility is present enhances striking power, accuracy of aim, and volume of fire, and – very importantly – the ability to escape to fight another day. Mobility on land, dependent at first on athletic warriors with endurance and foot speed, could be improved by use of the horse, and then by steam power or the internal combustion engine. On the sea it depended on sails filled by the wind or the hard labor of humans. As time progressed it would be enhanced by adding more humans to row the ship, better use of the wind, or improved hull design. It would be improved further by the use of steam power, the internal combustion engine, or nuclear-powered steam turbines.

Chapter 8 provides an example of the value of mobility on the sea. It describes defeat of the Spanish Armada by a diminutive force of brave and skillful British sailors operating ships with an innovative hull design and their leaders who used that technological advantage to great effect.

Chapter 9 carries the quest for mobility forward with the application of steam power on sea and on land, and use of the internal combustion engine on land to create a fearsome new weapon – the battle tank. It then introduces a new innovation, the airplane. It looks at early aerial combat and concludes with a brief treatment of jet and rocket propulsion.

Chapter 10 discusses the first of two overwhelming obstacles to getting the most out of new technologies: overcoming the Power of Prevailing Fashion.

Chapter 11 then discusses the difficulty of looking accurately into the future of technology.

Chapter 12 builds on the background provided by the first eleven chapters to discuss

Chapter 2 – Technology Makes Useful Weapons: Appropriate Doctrine Makes Those Weapons Useful

challenges existing today.

Chapter 13 describes a unique historical event – the first use of the nuclear weapon – its effectiveness in ending a bloody conflict and impact on U.S. defense doctrine up to the present time.

Note: These thirteen chapters are devoted to the kind of conflict commonly thought of when warfare is in view – conflict between nation states. The Appendix discusses the terrorism threat, notes the use of modern technology by terrorists, and raises the question of what technologies and methods of employment might best combat this threat.

War in its literal meaning is fighting ... Fighting has determined everything appertaining to arms and equipment, and these in turn modify the mode of fighting; there is, therefore a reciprocity of action between the two.

Clausewitz

CHAPTER 3

WARFARE: SOMETIMES LIMITED, SOMETIMES UNLIMITED.

Unlimited War in the Ancient World

Unlimited war was unfortunately common in the ancient world. One example is sufficient to show the effect that new tools of war can have in the hands of a ruthless aggressor. Iron in the hands of the Assyrians in the 9th Century BC enabled them to unleash a particularly brutal unlimited war on their neighbors. Under the command of the infamous Ashur-nasir-pal II, the Assyrian army conquered the Middle East using its ability to kill and by means of sheer terror. It burned opposing cities to the ground. Children were burned alive while their elders were either blinded and impaled upon stakes or left to die without limbs in fly-infested heaps. These campaigns of horror and intimidation were carried out deliberately, with little provocation, and the terror they caused was adequate to cause other besieged cities to capitulate without a fight.

Greek Limited War

By contrast, iron in the hands of the Greeks, until the first invasion of the Persians in 490 BC, facilitated a style of warfare that tended to be limited in objective. Greek knowledge of iron working had led not only to spear and sword but also to shield and body armor. The basic formation was the phalanx, the "germ of all future European

Chapter 3 – Warfare: Sometimes Limited, Sometimes Unlimited

military development', a block of armored infantrymen. Greek battles pitted phalanx against phalanx, in the open field, in the fall. Casualties were usually light. [1] During the intermittent conflicts between the Greeks themselves, food was scarce, and the primary objective of the Greek army in the field was often the autumn harvest of the antagonist.

In the case of wholesale invasion by the Persians the invaders contemplated unlimited war, The invading Persian general had pledged himself to lead the citizens of Athens "… away captives into Upper Asia, there to hear their doom from the lips of King Darius himself" [2]. The objective of the Greeks was more limited. It was simply to defeat the vastly larger Persian force, destroy it if possible, and remove the immediate threat to the city state of Athens. Although the Persian army was not destroyed it was driven back. Surviving Persians and reserves were then transported by sea to attack Athens. But the Athenian army had arrived back from Marathon and manned the city's defenses. The Persian fleet returned to Asia Minor ending this first invasion. A second would follow and key events in that one will be discussed in Chapter 4.

Conditions Tending to Limit War in the Middle Ages

Through the centuries, war has worn many faces, always brutal but sometimes less than others. In the Middle Ages, for example, the damage done by warfare in Europe was limited by two unique cultural manifestations: general restriction of war to the nobility, and the introduction of ransom. The first limited the numbers of participants involved and bound these participants by codes of honor which often served to reduce bloodshed. The second enabled a prisoner to secure his freedom, a ship owner to repurchase his ship from its captors, or even a city to prevent its sack. The right of ransom was recognized by law and not only reduced the fierceness of war but actually caused the practice of ransom to become a form of trade, reaching its peak in Italy in the 1400s when the quest for ransom "all but reduced fighting to a farce." [3]

Even where the art of ransom was not so highly developed, restriction of fighting in the Middle Ages to men of wealth and position, who could afford armor, and who observed codes of fighting prohibiting missile warfare, led to a pronounced reduction in casualties.

Many of the battles of this period were no more than shock skirmishes between small bodies of armored knights, in which individual combats were sought, to prove rather the worth of the fighter than his destructive capabilities. The object was to unhorse one's opponent rather than slay him. In short, battles were frequently little more than sharp-weapon tourneys.[4]

In England too, limits prevailed at various times, as shown by these ordinances of Henry V of England.

That no manner of man be so hardy as to go into any chamber of lodging where any woman lieth in childbed, in order to rob her, or pillage any goods belonging to her refreshing, nor make any affray whereby she or her child be in any disease or danger.

That no manner of man be so hardy to take from no man going to the plough and harrow, cart, horse, nor ox or nor other beast belonging to labour without payment and agreement.

That no manner of man beat down housing to burn, nor no apple trees, pear trees, nuts, nor not other trees bearing fruit.[5]

Rules such as these broke down both on the continent and in England, there particularly during the Thirty Years War from 1618 to 1648. Nevertheless, long after the age of chivalry had passed, as late as the eighteenth century, wars between European monarchs were viewed much as royal games. The pieces were highly drilled soldiers. Armies were kept small and "bloody encounters generally avoided." [6] This era was finally brought to an end by the industrial revolution and the war machines it produced, by the re-emergence of ideological war, and by the mass army.

The Impact of Mass Armies

The French revolution led to the earliest of the mass armies of Europe, first under the generals of the republic and then under Napoleon. The generals of the republic tended to use their armies recklessly, as those who failed were sometimes shot by the rulers of the Republic, and conscription, introduced in 1793, "gave them a glut of lives to squander lavishly". Napoleon also took ruthless advantage of the high flow of conscripted recruits in his attempts to conquer Europe. He is said to have bragged of

Chapter 3 – Warfare: Sometimes Limited, Sometimes Unlimited

his "income" of 200,000 young men a year, and stated on another occasion that he "did not take much heed of the lives of a million men." [7]

Modern War

The mass army, the machines of the industrial revolution, and the nationalism and interdependence of industrialized countries inevitably changed the face of war – leading to the horrifying carnage of what was called by those who lived to see it – "The World War." The great marshals of Europe did not accurately foresee the shape of the conflict. That was left to a banker, a Polish Jew, Monsieur I. S. Block, who published in 1897 a three volume work, "The War of the Future in Its Technical, Economic, and Political Relations" in which he accurately predicted the nature of the trench warfare which was to come during World War I.

> *At first there will be increased slaughter – increased slaughter on so terrible a scale as to render it impossible to get troops to push the battle to a decisive issue ... The war, instead of being a hand-to-hand contest in which the combatants measure their physical and moral superiority, will become a kind of stalemate, in which neither army being able to get at the other, both armies will be maintained in opposition to each other, threatening each other, but never able to deliver a final and decisive attack ... That is the future of war – not fighting, but famine, not the slaying of men, but the bankruptcy of nations and break-up of the whole social organization ... Everybody will be entrenched in the next war. It will be a great war of entrenchments. The spade will be as indispensable to a soldier as his rifle. All wars will necessity partake of the character of siege operations ...Soldiers may fight as they please, the ultimate decision is in the hands of famine ...*[8]

Not only had the mass army returned to warfare, but – with the advent of Hitler's "Lightning War" in 1939, the return of the time-honored sneak attack in 1941, and the introduction of mass aerial bombardment in World War II – civilian populations again became unwilling participants in warfare, as they had been ages before during the campaigns of the Assyrians and the sack of Constantinople. The strategic nuclear weapon then completed the circle.

Innovation in Weapon Systems

Destruction and the threat of destruction of civilian populations once again became tools of would-be aggressors, just as they had been tools of the Assyrian kings. In contrast to the first Persian campaign against Greece, where Greek superiority in weapons technology, tactics, and the character of the individual soldier protected an entire population from death or Persian enslavement, protection of societies today against the threat of wholesale destruction is considered by many to be impossible.

It was the armament of the Macedonians coupled with the genius of Alexander the Great which, by overthrowing the Persian Empire in the fourth century S.C. led to the emergence of the Hellenistic cultural period, which ... profoundly influenced Rome's destiny. Again ... five centuries later, it was the armament of the Goths which was one of the main factors in the extinction of the Western Empire. Later still, that it was the armament of the Eastern (Byzantine) Empire which secured its existence until 1454, when yet a new weapon not only proclaimed its end, also the end of medieval civilization throughout Europe.

J. C. F. Fuller

CHAPTER 4

THE SEARCH FOR RANGE: EARLY WEAPONS

Early Use of the Bow

The arrow and bow combination, a weapon extending the range at which damage could be inflicted upon an opponent, was used prior to the existence of coherent records of history. In Egypt, excavations have uncovered a group of some 80 painted soldiers in the tomb of Mesehti at Asyut. One half of the foot high statues represent Nubian archers – from the southern reaches of the Nile above the first cataract – complete with bows and arrows; the balance are Egyptian spearmen, carrying large shields. [1]

Some idea of the value of the bow in early warfare can be gained from the available descriptions of conflicts between Greek and Persian forces in the fifth century B.C. These descriptions reveal that weapons are indeed important, but only with tactics –

accidental or planned – appropriate to these weapons and to the situation.

In the Greek city state every free born citizen served in the militia, whenever needed, over the greater part of his active life. For example, Socrates fought a campaign when he was forty seven. [2] The backbone of the militia was the hoplite, a pikeman who carried, in addition to his six foot spear, a short sword and shield, and wore helmet, cuirass, and greaves. It was not until the later rise of Phlilip of Macedon that strong supporting forces of cavalry and archers would be added to what was basically a Greek army of infantrymen.

In the Second Persian War, in 480 b.c. Xerxes, the son of Darius the Great, began a large scale invasion of Greece. Historians tell us that he moved a force of some 360,000 men over a bridge of boats across the Hellespont and advanced upon a Greek confederation whose plan of defense was the outcome of much bickering and compromise. A group of outnumbered Greek infantry with a Spartan core made a stand at Thermopylae. The Persian infantry fearful of their leaders but more fearful of the Greek spears and swords, refused to close with the Greek hoplites; but the Spartans and their companions were eventually "worn down by a hail of arrows from both front and rear." [3] Arthur Birnie described Thermopylae as, "a military disaster which the Greeks tried to camouflage by dwelling on the heroism of the defeated." [4]

After destroying the Greek force at Thermopylae, capturing Athens, and fighting an indecisive but damaging sea battle with the Athenian navy, Xerxes returned to Asia, leaving behind his general, Mardonius, with a force of 100,000 men to complete the conquest of Greece. However, after Spartan forces agreed to join those from Athens, Mardonius retreated through the Cithaeron–Parnes range and camped. The pursuing Greek force, carried away by its success in a minor encounter with the Persian cavalry on rough ground that had favored the Greek infantry, decided to press an attack on the plain of Plataeu. The Greek infantry formed behind a ridge and prepared to attack. However, the Persian leader used his cavalry to circle to the rear and destroy the spring used by the Greek army for a water supply, then sent his cavalry against the Greek infantry. Repeating a Persian tactic that had been successful before, archers stayed out of reach of the Greek spears and for two days, "from a safe distance poured a hail of arrows and spears on the devoted hoplites." [5]

The tormented Greeks attempted a night retreat, but the move was uncoordinated, and when morning light arrived some were caught retreating across the plain and

Chapter 4 – The Search For Range: Early Weapons

again subjected to a hail of arrows.

At this point, a tactical blunder by the Persian general saved the Greek army. Rather than continue to take advantage of the range of action of the bow, which was taking a toll of the Greeks with only small Persian loss, Margonius elected to press the attack upon the Greek force with division upon division of Persian footmen. These unfortunates, upon encountering the hedge of spears of the Phalanx, piled up one upon the other until they could neither move nor retreat. The desperate Greeks charged, and in the hand-to-hand combat which followed, the lightly armored Persians proved to be no match for the fierce hoplites who drove them from the field in wild flight.

Plataea is an example of how a first-rate strategic plan can be ruined through a tactical error. On the morning of the battle Mardonius had his enemy beaten. He has only to continue his harassing tactics to see the scattered Greek forces drift back through the passes and dissolve into fragments. But through over-eagerness to insure their destruction his allowed his men to be trapped into a hand-to-hand encounter with superior troops, and he lost everything. "Desperate courage won the stakes ... By sheer hard fighting the hoplites won..." [6]

In 331 B.C. the bow also played a part in the opening phase of the decisive battle of Alexander's war against the Persian Empire at Arbella, near the ruins of ancient Nineveh. Prior to this battle, Darius III, the emperor of a now-unstable Persian Empire, had used Greek mercenaries to strengthen his infantry, but by this time most had been killed or had deserted. Hastily trained Persians took their place but were such unsatisfactory replacements, that to strengthen his infantry force, Darius interspersed cavalry units with it, and protected its front by a row of scythed chariots. However, as the battle commenced, Alexander put his archers and javelin men opposite the chariots and they quickly shot down the drivers and chased the terrified horses from the field before the Persian army could close. Alexander's trained, well-armed, and well armored force that day won what has later come to be regarded as one of the most important battles in world history.

There were other incidents that contributed to the victory. For example, after the armies came together, the surge of battle opened up a gap in the Greek front. A body of Persian and Indian cavalry charged into the gap in perfect position to attack the Greek phalanx from the rear. But instead they rode off to plunder the Greek baggage train and little damage was done. Later, Darius fled from the field and when word of that event circulated, the Persian army began to lose heart. Eventually, it broke and fled, and was slaughtered by the pursuing Greeks. [7]

Byzantine Use of the Bow and Other Weapons

Constantinople was founded in 330 A.D. by the Roman emperor Constantine, who set up a senate, as in the West, and a duplicate of the imperial bureaucracy. The Eastern Roman Empire combined Greek culture and a state religion into a fervent nationalism that helped the empire survive for almost a thousand years after the fall of Rome. Rich agricultural regions enabled the growth of cities which contained almost half its population (a high proportion for that age) carrying out production and trade of textiles, glass, metal ware, and other industrial goods, enhancing the wealth of the empire.

In certain respects it is possible that the situation of the Byzantine empire, through much of its existence, might be considered somewhat analogous to that of the United States in the 20th and early 21st centuries. It was less interested in conquest than in keeping what it had. A pronounced difference however was that the empire lay in the path of a multitude of invaders and its wealth was a permanent temptation to outsiders.

> *Throughout its long history, the empire was called upon to withstand countless attacks from such disparate enemies as the marauding Huns, the migrant Slavs and Germans, the fanatical Arabs, the mercenary Persians, and later, from equally rapacious Western European enemies, the Normans and the Crusaders.*[8]

The military power and internal strength of Byzantium is apparent in view of both its endurance and military campaigns conducted in Italy in the 6th century. [9] Although the Western Roman Empire had been dismembered by Gothic warlords in the 5th Century, the Eastern empire continued as a military power until the end of the first millennium, even recapturing much of the lost Roman lands in Southern and Central Italy from the Ostrogoths during the reign of Justinian. Even after Italy was lost, as long as the empire stood, Byzantine military force blocked what would otherwise have been the major thrust of Arab armies into Europe.

Chapter 4 – The Search For Range: Early Weapons

The Arabs conquered Spain in 717, and Arab raiding parties, filtering through the Pyrenees, probed into Southern France. It was one of those parties which was defeated by a Frankish army in the desperate battle at Tours in 732. The fierce struggle of Charles Martell with one tiny tentacle of Arab power takes on a somewhat different significance from that usually attributed to it when it is remembered that what really stood between the Arabs and the more direct approach to Western Europe was the double bastion of Byzantine land and naval forces. It need hardly be added that when the bulwark of Byzantium disappeared, Moslem armies striking through the Balkans were able to reach the gates of Vienna in 1529.[9]

Time took its toll of even this well-knit society however. After the end of the first millennium, during the reign of Basil II, the empire seems to have begun a precipitous slide downhill. In his eagerness to get riches to support an inefficient and swollen government bureaucracy, Basil II conquered, and then left defenseless against the Turks, the Christian kingdom of Armenia, which was a natural ally of Constantinople and should have been treated as such. This exposed the empire to attack from that region. Following the death of Basil the Second, rulership of the empire passed, as described by Fuller, "... by intrigue, scandal, and licentiousness" through a series of degenerate rulers, whose reigns saw a steady decline in the strength, character, and vigor of the entire society.[10]

The Eastern Roman View of War

The Byzantine approach to warfare was remarkably different from that of the feudal society in Western Europe. Western war depended on courage, strength, and weight of forces. In Byzantium, the continued existence of the empire depended completely on the existence of its armed forces; consequently, the study of war, ignored in more complacent societies, occupied the best minds. The underlying tone of the resulting Byzantine writings on warfare was caution. Fighting was to be avoided if at all possible. It was a rule never to use force if money would do. Because this policy was backed by military power, it did not weaken the country as simple appeasement might have done.[11]

If fighting was necessary, the most desirable victory was the one requiring the least loss of men and materials. The Byzantines did not view peace as normal and war as abnormal. Rather, they looked upon war as the inevitable result of interactions of groups of people, and disdained any military adventures which would weaken it when the next war came along.

The Byzantines were acutely concerned with the preservation of scarce resources, particularly human life. Theirs was the only state of that time to pay any attention to treatment of the wounded. Their military textbooks show that the army contained an organized ambulance corps of bearers and surgeons. Bearers were paid a bonus for every casualty carried in from the field of battle.[12]

The Byzantine Horse-Archer

Since Byzantine armies were often outnumbered, the society gave appropriate attention to the development of superior weapons and tactics. The representative Eastern Roman soldier was the horse archer. Like the individual Western knight, he was clothed in mail, but there the similarity ended. He had the typical mobility and shock power of the cavalry soldier but also a skill with bow that made him, "the most versatile and one of the most effective cavalry soldiers in the history of warfare." [13] The horse archer was supported by cavalry-lancers and heavily armored infantry, but the highly trained mounted bowman was the difference between the imperial army and opposing armies for much of the duration of the empire. In the words of Justinian's Cavalry General Belisarius (505-565):

> *I found that the chief difference between them (the Goths) and us was that our Roman horse and our Hunnish Foederati are all capital horse-bowmem, while the enemy has hardly any knowledge whatever of archery. For the Gothic knights use lance and sword alone, while their bowmen on foot are always drawn up to the rear under cover of the heavy squadrons. So their horsemen are no good till the battle comes to close quarters, and can easily be shot down while standing in battle array before the moment of contact arrives. Their foot archers, on the other hand, will never dare to advance against cavalry, and keep too far back.[14]*

Effective use of the bow meshed perfectly with the Eastern empire's view of war as a practical, and not a heroic affair. To sacrifice lives and gain by valor would could more cheaply be achieved by cunning was considered, "the worst of bad general-ship." [15]

An example is furnished by the battle of Taginae, won in 552 by the eunuch, Narses, over the Gothic king, Totila, which Fuller calls an exact forerunner of Crecy. Narses dismounted 10,000 of his knights, keeping a smaller remainder mounted in reserve. On each wing he put some 2,000 archers. The Gothic army charged the dismounted knights, intent on wiping them out, but was decimated by the combined force of Byzantine archers, and driven off the field by the mounted reserve.

When the armies of Eastern Rome were first attacked by Moslem fanatics in the 7th Century, they faced clouds of Arab cavalry thrown unorganized into battle, and beat them. When the war resumed after a respite of years, the Arabs had learned from the Eastern Romans. They had armored their cavalry and had trained their warriors to attack in a solid line. Although the Byzantines had a slight edge in numbers, they were not willing to fight a war of exhaustion with a similarly equipped army. Consequently, they had used the period of peace to prepare for this next war, developing cavalry tactics similar to the infantry tactics of the old Roman legions. Just as Roman use of the maniple had called for remarkable courage and discipline, so did the Byzantine use of blocks of cavalry called "banda" in successive waves. These wave attacks proved successful in disrupting the Arab ranks, making them prey for the imperial horse archers and foot archers.

Byzantine Naval Superiority

The major sea battles in the continuing conflict between the Byzantine fleet and Arab fleets took place along primary trade routes and near land bases in the Mediterranean. Byzantine ships were descendents of the galleys of the Western Roman Empire and were similarly propelled by oars. This led to a major limiting factor with these ships – range. Storage space aboard ship was limited and large quantities of water were needed for the relatively large crews, particularly those individuals manning the oars. A small galley would have to replenish its water supply every two and a half days. The larger vessels would have to put into port on the average of once a day and we can imagine that a large fleet would be very difficult to maneuver in and out of port to reload water. In the 10th century for example, a Byzantine fleet of some 177 ships of various sizes assembled for an attempted invasion of the island of Crete. This involved a force of more than 40,000 men including soldiers and Marines on board the ships. In this case supply boats had to be used to ferry water out to the fleet.

In a burst of perhaps an hour or less the ships might move at something like 10 knots but sustained speed was much lower. Under sail in the most favorable conditions it was 3 to 4 knots.[16]

Arab ships tended to be copies of the Byzantines ships. They were built when the Arabs recognized that control of sea lanes was necessary for trade and as Arab armies overwhelmed ports along the south of the Mediterranean where shipbuilding skills were relatively well developed.

Throughout most of its history, the Eastern Empire was a trading nation. Its merchant fleet required safe access to Mediterranean trading centers and in turn the merchant fleet supplied manpower to the navy, which maintained control over the routes to foreign commercial centers. Naval bases and yards were established at Carthage, Acre, and Alexandria. The central base, with harbor, dockyard, and arsenal, was at Constantinople. Besides a trained manpower base and extensive supplies of naval stores including timber and iron, the Byzantines also possessed an awesome naval weapon, the mysterious Greek Fire, or wet fire, which was expelled from brass tubes using pumped sea water.

The first use of this weapon apparently occurred during the reign of Constantine IV (668–685), when Moawiya, the first of the Omayyad Caliphs, attempted to capture Constantinople by forcing his fleet up the straits. He was partially successful, his fleet capturing the strategic city of Cyzicus and using it as a base to blockade the Bosphorus and attack Constantinople itself. After enduring this situation for five years the Byzantines were able to gain a tactical surprise with this revolutionary new weapon and destroy the Arab fleet. Thereafter, Greek Fire continued to give the imperial fleet a technological and psychological advantage over Arab fleets that attempted twice in the 8th century to force their way through the Dardanelles.

The Byzantines regarded themselves as Romans, and did not refer to their weapon as "Greek Fire." This name was given to it later by crusaders from the West. Fuller refers to Greek Fire as one of the few true master weapons in the history of mankind.[17] The formula for Greek Fire was a carefully guarded secret. Today, although there is speculation about its make up, precisely what it was is unknown. When Arab fleets finally took control of the Western Mediterranean in the early part of the 10th Century, it was apparently partly due not only to the decline of the imperial fleet in size and quality of personnel, but also to the fact that the Arabs had "narrowed the technological gap" by development of an incendiary weapon somewhat similar to Greek fire.[18]

The fire was a flammable liquid sprayed out of the nozzle of a metal-clad tube. Its origination is sometimes attributed to a Syrian who lived in Constantinople but its precise origin is unknown. As noted above the exact ingredients used are likewise unknown. There is speculation that it may have been formed from a combination of naphtha distilled from pools of crude oil found in present-day Azerbaijan. This may have been mixed with equally flammable stabilizing ingredients like wax, quicklime, sulfur, and turpentine, and then distilled to create a flammable but stable liquid. It has been suggested that a commander of one of the regional Byzantine fleets defected to the Arabs in the middle of the ninth century, taking with him knowledge of Greek Fire.

The incendiary liquid was housed in a sealed tank which was heated to near boiling point by means of a furnace that was served by one or two sets of hand operated bellows and when the device was ready to be used crewmen would pump air into the tank by means of a two-handed pump.

When the pressure was sufficiently high inside the tank the liquid would be forced to a tube. As it was ejected out of the nozzle it was ignited. Ranges of around 12 to 15 meters were possible – more than enough to cover the gap between galleys. Greek fire burned on everything it hit on the enemy ship and even on the surface of the sea itself. It is not surprising it was called infernal fire. Because of its high strategic value and the secrecy surrounding it, its required projectors were fitted only to the ships of the Imperial fleet. [19]

The Short Bow in Europe

Although the short bow would eventually decline in relative importance due to the emergence of the crossbow, the English longbow, and subsequent weapon developments, its use in the 11th Century in the form of the Norman short bow provides what Preston and Wise call "... one of the clearest examples of the influence of armament upon history." [20]

The Battle of Hastings

Normandy received its name from the fierce Northmen who populated Brittany following its capture by the Viking leader, Rolf, around 900 A.D. Some 150 years later, one of Rolf's descendants, deprived of a claim to the English throne, invaded England to take the crown from Harold Godwinson. The invasion fleet of this descendant of Rolf, William the Conqueror, landed unopposed in Sussex, and his army advanced against Harold's hastily gathered force of Anglo-Saxon infantry on October 16, 1066.

The defenders relied for protection on a hedge of shields, held by the soldiers standing shoulder to shoulder. Their primary weapon was a two-handled battle axe, with which Saxon infantrymen on foot could sever the head from the mount of an opposing horseman. This was effectively demonstrated when repeated assaults by the Norman cavalry against the wall of shields were beaten back. Realizing that a frontal attack could not break through the wall of shields, William commanded the Norman bowmen to aim, not directly at the opposing infantry, but to shoot their arrows into the air "... that their cloud might spread darkness over the enemy's ranks." [21] The effect was devastating. Many infantrymen were slain immediately. Others tried to protect their heads with their shields and could not use their battle axes to full effect. With the shield wall disrupted, the Norman horsemen easily broke through and destroyed the Anglo-Saxon infantry.

While many historians have referred to Hastings as a victory of cavalry over infantry, the range of the missiles from Norman bows was the critical factor. However protection for the source of those missiles was necessary. Fuller points out that if the Norman archers had not been supported by cavalry, Harold's infantry could have driven them off the field. What prevailed was a combination of forces, with the range of action of the bow making it the dominant weapon. Despite this shocking display of the power of the bow – the importance of its range – it would be three centuries before English archers would use the bow to revolutionize European battle tactics, and even then Western chivalry on the continent would refuse to adopt it. Fuller attributes this Western opposition to the bow to "... the hypothesis that missile fighting was as abhorrent to western military ideals as gas warfare is today." [22]

Chapter 4 – The Search For Range: Early Weapons

The Crossbow

The fearsome horse archer of the Eastern Roman Empire had used a conventional short bow. It appears that the Byzantines were unfamiliar with the crossbow until they saw it in the hands of Western Europeans. Since this did not occur until 250 years after the Viking raids on Northern France, the weapon may have been introduced into Europe by these marauders, who in turn may have learned of it from the Saracens, who had probably imported it from the East.[23]

Although it was more powerful than the conventional short bow, its low rate of fire made the crossbow more suitable for siege operations than for use in the open field, but its power and range eventually caused it to be adopted throughout Europe. By 1199 A.D. it had apparently spread throughout Italy and Southern Europe. Such widespread use is evidenced by the issuance in that year of a Papal Bull terming the crossbow too barbarous for one Christian to turn against another. Some 50 years later, after Richard I of England and Phillip II of France had both equipped their forces with it, Pope Innocent III repeated the church's position. However, it continued in use in Europe, and, in heavy steel form, was used by Genoese mercenaries in the employ of the French at the battle of Crecy in 1346.

The Longbow In the Hundred Year's War

The slow rate of fire of the crossbow and its reduced effectiveness in wet weather due to variable performance of the stringing, caused the English to prefer another weapon, of Celtic origin, the longbow. The English formally adopted the longbow for military purposes in 1285, when Edward I declared the longbow the basic weapon of the freemen of the shire levy.

The English yew from which the most powerful longbow was made was in such great demand in England that English bow makers were forced by government edict to make four bows of alternative materials (witch-hazel, ash, or elm) for every one made of yew, and restrictions of age and means were placed on who could even shoot a yew bow. For example, no person under 17 could shoot a yew bow unless he possessed a personal wealth of approximately 520 shillings or was the son of parents "... having an estate of ten pounds per annum." [24]

Since the power of a bow is proportional not only to the tension of its string, but also to the distance through which it acts, the longbow was a powerful weapon in the hands of skilled English freemen. Not only was it extremely powerful, it was light, as were its arrows. The crossbow, with its steel bolts, was heavy and unwieldy. An English bowman could fire a dozen or more arrows while a cross bow was being wound up. Its power was awesome. In the conflict between Henry II (1154-1189) and the men of South Wales, Gerald de Barry, the King's chaplain, saw a Welsh archer drive an arrow through a thick (four fingers) oak door. More unhappily, in the same conflict, one of the King's knights was pinned to his slain horse by a Welsh arrow which penetrated, in turn, armor, thigh, saddle, and mount.

Some authorities say that the longbow was slightly outranged by the crossbow and that its other characteristics provided its advantage. Others insist that the longbow had a longer range than even the highly refined steel crossbows carried by the Genoese at Crecy.

Crecy, Potiers, and Agincourt

The power of the English longbow was demonstrated repeatedly during the Hundred Year's War with France. The first and possibly most striking event was the Battle of Crecy, fought on August 26, 1346. The opposing French force, some 60,000 strong, under Phillip of Valois, was feudal in nature. It included 8,000 mounted knights and 5,000 Genoese crossbowmen. The French despised infantry and thought reliance on missile weapons was unknightly. The British force, led by Edward II, was semi-national, out-numbered three to one (only 800 knights, 8,000 infantry, and 11,000 long bowmen) and thought French ideas of chivalry absurd.[25]

The French leader sent his hired Geonese crossbowmen forward against the English bowmen, but they were apparently outranged, and, in any event, were decimated by the arrows from English longbows. The French knights suspected treason and charged forward through the Genoese, killing many more, only to be driven back themselves by the English bowmen. The French knights then launched charge after charge against the English infantry which was supported by the English archers, only to be beaten back each time. When night fell a third of the French force had been killed, the majority by arrows from the English longbow.[26]

Chapter 4 – The Search For Range: Early Weapons

The French refused to believe that the longbow in the hands of English commoners could be decisive. At Poitiers ten years later, King John of France apparently believing he was copying the successful English tactics, dismounted his knights for the attack, and launched them on foot against the English infantry. Long before they reached close quarters they were demoralized by the hail of English arrows. Attack from the side by a small infantry contingent and a small reserve cavalry force completed the defeat of the French force. Among those captured were the French king and his son.

It was not until 1370 that geography, internal troubles in England, and the appointment of a wiser constable of France turned the tide, at least temporarily, in favor of France. The French king found in Bertrand de Guesclin a leader who recognized that it was not only suicidal but unnecessary to fight the English in the open field. This French general sent the nobles, anxious for another battle with the English, back to their castles where they were safe from English arrows, and used professional soldiers of free companies to fight a war of harassment, ambushes, and slow sieges to reduce, little by little, the British holdings in France.

After his passing, however, the French nobility asserted itself again and at the Battle of Agincourt in 1415 "… the chivalry of France showed that it had forgotten nothing and remembered nothing." [27] The English force under Henry V numbered only 6,000 men. Five thousand were archers. This small force was stationed with its flanks protected by heavy woods and its front by almost a mile of wet, freshly plowed fields. The French constable, a pupil of de Guesclin, would not attack, and planned to starve the English out of their favorable position. But the French nobility, confident once again in a three-to-one numerical superiority, overrode his decision and attacked.

Astonishingly, they repeated the tactics used at Poitiers. Dismounting and sending their horses to the rear, the French knights in full armor plodded through the ankle deep mud, intent upon destroying the English commoners armed with longbows. The slow-moving force received volley after volley of arrows from the English longbows. Those who survived the hail of arrows were exhausted when they reached the English force, where the unarmored English yeomen abandoned their bows for axes and swords and killed the rest. English losses at Poitiers are believed to be at maximum a few hundred, while more than 4,000 French nobles and knights were killed.

Eventual English Defeat In France

Despite the brutal defeats inflicted on the French, several factors contributed to eventual expulsion of the English forces from France. These included English unrest at home because of higher and higher taxes, the strain of a nation of two million people trying to sustain a long conflict against a nation of 16 million, and the rise of the French masses, led by Joan of Arc, against the occupying British forces. But the most decisive factor seems to have been the adoption of a new weapon by the French: cannon.

In a three year period, from 1450 to 1453, a single French artillery train methodically assaulted, one by one, the English castles which had controlled large portions of Normandy. And at Formingy in 1450, an English army, arrayed as at Potiers and Agincourt, was cut down by French artillery placed on each flank. The outnumbered English archers broke ranks, not to flee, but to try to capture the French guns, but were defeated by French superiority in numbers in hand to hand fighting.

For over one hundred years, from 1346 to 1450, the English longbow had brought the finest French cavalry to a standstill, but then was suddenly driven from the field by a new weapon with even greater reach and destructive power. The age of gunpowder had begun.

Man is eternally inventive. No sooner does he bring a system of destruction to perfection than the constant technical factor – the urge to eliminate the danger he has created – under the whip of disaster compels him to seek yet another.

J.C.F. Fuller

CHAPTER 5

THE SEARCH FOR RANGE: THE AGE OF GUNPOWDER

Catapults and Gunpowder

Prior to the invention of gunpowder, heavy projectiles could be propelled at an enemy only by use of catapults. The art of constructing and using various forms of such catapults, known in the ancient world and mentioned in fact in the book of Chronicles, reached a peak in the army of Alexander the Great. These devices continued in use in the Middle Ages, and, although they were generally used for throwing heavy stones, other projectiles were sometimes used. Cleator, for example, tells of the use of catapults to throw dead horses over castle walls, and at the siege of Carolstein in 1422, to shower down upon the defenders the contents of two thousand cartloads of manure – an early, if unintentional, instance of germ warfare.[1]

The first invention of gunpowder is generally ascribed to the Chinese. A Chinese treatise believed to date back to the year 1044 gives a recipe for an explosive mixture containing sulfur, charcoal, and saltpeter, the essentially ingredients for the explosive mixture that eventually found its way into primitive cannon.

The earliest known document mentioning cannon is in Arabic and is dated 1304. In the illuminated manuscript of 1326, now found in Christ Church, Oxford, there is a picture of the earliest known cannon, the "dart-throwing vase." [2] In 1339 a multi-barrel weapon was used by Edward III of England against French forces.

Early Cannon

Gunpowder found its earliest use in artillery, rather than small arms. This was no doubt due to the heavy, cumbersome nature of early hand-held gunpowder devices, which could not at first compete with hand powered weapons. Cleator describes the early development of cannon:[3] For the first century after their appearance in Europe (roughly 1326-1426) these primitive devices had little effect on the outcome of wars. The first cannon were small, fixed to wooden beams attached to wooden frameworks, and were prone to explode – as dangerous to the users as to their opponents. Increases in size and power came at a sacrifice of mobility due to excessive weight of cannon and supporting framework. Most early guns were of copper, brass, or bronze, with brazed or riveted seams. By 1400 small pieces could be produced by pouring molten metal into a mold. Very large cannon were made by packing wrought iron rods longitudinally around a mandrel, and then forcing red hot bands of iron over the rods so that when they cooled and shrank, they packed the rods tightly together. The resulting tube was open at both ends. The makers then cast or forged a separate powder chamber with a tapered end, and wedged or screwed it into place at the breech end. These early guns were muzzle loaders without gas-tight joints between powder chamber and barrel, and suffered from the constant danger of the powder chamber being blown from its crude attachment to the barrel.

Early mortars, although short and squat and therefore generally easier to cast in one piece, suffered from a unique operational problem of their own. The user was required to utilize two linstocks, one to light the fuse of the cast metal, hollow, powder-filled bomb, and one to light the powder at the touch-hold of the mortar itself. "A dual operation accompanied by the threat of imminent disaster, if for any reason the latter misfired." [4] The now-obvious answer to this problem, but not at all obvious to users at the time, was to let the bomb fuse be ignited by the flame from the burning propelling charge. But this procedure was not adopted until more than 100 years after the introduction of early mortars.

Chapter 5 - The Search For Range: The Age Of Gunpowder

Improvements to the First Cannon

Cast Iron

The first cast iron cannon were made in Western Germany about 1400. The practice reached England 100 years later when the English monarchy imported two foreign craftsmen, Peter Van Collen and Peter Baud, who also developed hollow explosive cannon balls. (Whether or not this was the earliest construction of such cannon balls is uncertain. The Venetians may have built such projectiles in 1376, and the Dutch certainly did so in 1508.) [5] At any rate it was the Dutch who thought of lighting the fuse of the explosive charge with the flame of the cannon's discharge. Once developments reached this stage, proliferation of models followed. Various manufacturers began to build cannon of individualistic sizes and designs, and in England by 1603 the official ordinance list showed the existence of 16 different pieces, ranging in size from one to eight and one half inches in bore, with shot ranging from one half pound up to 66 pounds.

Light Weight Cannon

Mobility in the field was a continuing concern. Two developments permitted designers to lighten cannon by reduction in barrel length: (1) improvements in gun powder, and (2) more uniform shot sizes which reduced power losses from loose-fitting shot. The search for lighter weight also led to unusual developments such as the use of wooden artillery (Henry VIII at the siege of Boulogne in 1544) and the "Kalter" or leathergun of Gustavus II Adolphus of Sweden (1594-1632). The actual construction of these latter pieces is uncertain. It appears that they were thin copper tubes, either encased in hide, or banded with iron and wound with rope. Their advantage was light weight (about 90 pounds) but their disadvantage was a tendency to overheat and cause pre-ignition of the powder charge. [6]

Ballistics Developments

More effective use of these early firearms came not only from innovations in the primitive fields of metallurgy and weapons design, but from the theorist who could better explain how these weapons worked and how to use them. In the early 1500's the mathematician Niccolo Fontana Tartaglia, combined a plumb line and graduated scale to make a "gunner's quadrant" to measure the elevation of a field gun. Although Tartaglia's knowledge of ballistics did not match today's state of the art, he was able to make improvements in the theory of that day which believed, for example, that a cannon ball fired from a muzzle continued in a straight line to some point in space where it suddenly changed direction, and fell straight to earth. [7]

Between 1400 and 1500, the principal use of artillery was to knock down castle and city walls. An important exception was the Battle of Formigny, in 1450, described earlier. This battle, and the French re-conquest of Normandy which accompanied it, ended the Hundred Years War. Using their primitive cannon to blast the English out of previously impregnable castles, the French carried out 60 successful siege operations in less than a year and a half. [8]

The Siege of Constantinople

If the age of gunpowder did not begin at Formigny in 1450, it began at the siege of Constantinople in 1453. The cannon used against Constantinople by Mahomet II were cast of bronze by a Hungarian renegade cannon maker.[9] There were thirteen "great bombards" and fifty-six smaller cannon. The large pieces had barrels 60 feet in length and required 60 oxen to pull then across country, while 200 men were needed to level a road for their progress and 200 more to march alongside to keep them in position. [10] Ramps of earth were used as firing platforms. They did not use iron shot but fired large stones 30 inches in diameter, weighing approximately 1200 pounds. Loading time was two hours, so each cannon fired only seven times per day. When those Turkish cannon finally appeared outside the walls of Constantinople, a society which had endured for over a thousand years through careful attention to defense, provision of superior weapons, and economic use of its military forces was about to suffer the consequence of decline and neglect of its once powerful defenses. The bombardment began on April 12, 1453. By May 29 an opening had been made in the walls, and the Turks overran the city.

Had the inhabitants of Constantinople known what awaited them and possessed initiative and armed forces of the quality of the old empire, they could possibly have prevented the siege. Turkish progress across country with the cannon was slow,

Chapter 5 – The Search For Range: The Age Of Gunpowder

approximately three miles per day, and nearly two months had been required to drag them across county and place them opposite the walls of the city.

TABLE 5-1

A Partial List of Early Gunpowder Devices

Hand Grenades	1382
Smoke Balls	1405
Time Match	1405
Case Shot	1410
Matchlock or "Arquebus"	1450
Bronze Explosive Shell	1463
Explosive Bombs	1470
Wheeled Gun Carriage	1470 approx.
Pistol	1493
Incendiary Shell	1487
Rifling	1520
Wheel Lock and Spanish Musket	1521
Improved Hand Grenades	1536
Wheel Lock Pistol	1543
Paper Cartridges	1560
A type of shrapnel shell	1573
Hot Shot	1575
Fixed Cartridges (powder and ball in one)	1590
Rifled pistols	1592
Percussion Fuse	1596

Source: J. C. F. Fuller, <u>Armaments and History</u>, P. 81.

French Cannon in Italy – The Doom of Fortifications?

In the second half of the 1800's the English used primitive cannon in the bitter War of the Roses (1454-1482). This war caused widespread damage and was followed by a corresponding widespread and intense English dislike for the soldier, which left England without an army for 150 years, while nations on the continent surged ahead with land military developments. The most startling development during this period was exhibited in the invasion of Italy in 1494 by Charles VIII of France.

Only 79 years before, French knights had slogged through the mud at Agincourt and been cut down by the English longbow. Now, in 1494, a new weapon, the cannon, turned the French army into a nearly irresistible force on the continent. Italy had been getting along very well, with its internal warfare between cities handled by mercenary armies, the "Condottieri", who substituted cautious operations and ransom for serious fighting. When the pendulum swung back toward unlimited war, they were mentally and materially unable to cope with the change. Their experience in limited war, and their technology, were no match for the French:

> *The barbarians from across the Alps fought to kill ... In 1495, at Fornovo when the French were fighting their way back, the Italian men-at-arms were routed by the French mounted "gendarmerie" and the French artillery. The next day, the Italian commander, the Marquis of Mantua, coming to ransom his friends and relatives found to his horror that they had all been killed in battle. A new age had dawned.[11]*

The two armies provided a contrast. Not only in purpose, but in technology as well. The French cannon were a world apart from the clumsy field pieces of the Condottieri, which had to be dragged laboriously across country by large teams of oxen. Charles VIII had planned carefully for his invasion of Italy. He decreed that the barrels of his cannon could be no more than eight feet long, and also provided wheeled carriages for mobility, and trunions so that their muzzles could be readily elevated or depressed. As a result, the French horse-drawn cannon could keep up with infantry on the march and quickly bring their fire to bear on Italian fortifications. The success of French artillery against previously impregnable Italian fortifications was so startling that a general belief arose, "dictated by fright rather than reason"[12] that all fortifications were worthless.

Artillery In the Open Field – Temporarily a Master Weapon

It was not only in the destruction of fortifications that the cannon became a weapon of mass destruction. It was used by both combatants in the open field at the costly Battle of Revenna in 1512 between French forces and the army of the Holy League.

Chapter 5 – The Search For Range: The Age Of Gunpowder

Artillery caused large-scale slaughter on both sides, and the battle ended only when the French charged Spanish artillery positions and drove the gunners away in hand to hand fighting. Subsequently, at Marignano, a Swiss pike column was decimated by artillery, and the remnants driven off the field by cavalry, although historians say that the carnage caused by the artillery fire was so complete that the cavalry charge was unnecessary. [13] (Note: The pike had previously shown itself capable of stopping sword-bearing cavalry and had enjoyed its own brief period of being a near-dominant weapon.)

In an attempt to reduce the losses from artillery, battlefield commanders began more and more to dig in. And defenders found the new weapon was not irresistible after all. Wet ditches, plus ramparts, and bastions covered by earth and mounting heavy guns of their own replaced castle, moat, wall and tower. In 1519 the Emperor Maximilian brought a siege train, many times more powerful that that carried to Italy by the French in 1496, to the outskirts of Padua. His artillery was completely defeated. The weapon that had destroyed Italian castles or driven their terrified occupants to surrender only 25 years earlier was defeated by a new emphasis on the defensive. From 1521 on, successful sieges were few.[14]

Uneven Progress In Artillery

When developments in heavy artillery came to a temporary standstill, technological progress in hand-held arms caught much of the tactician's attention in the 1600's. Then, between 1700 and the American Revolution, the state of technology of infantry arms stood still, while considerable strides were again made in the development of artillery. [15] This was the period of Benjamin Robins and the ballistic pendulum. In the late 1700's artillery development slowed down once more and was again overshadowed by small arms developments, although isolated artillery developments did occur, such as during the siege of Gibraltar, when the British developed a new explosive projectile, "Mercier's operative gun shell" to damage attacking ships.[16]

Small arms development continued to hold the attention of military authorities in the early and mid 1800's, until revolutionary new developments in artillery technology, breech-loading and other improvements, again made artillery, if only temporarily, the dominant weapon at the opening of the 20th Century.

Breech-Loading Cannon and The Franco-Prussian War

There had been a continuing interest in the idea of breech-loading artillery pieces. Earlier cannon, as mentioned previously, were constructed with a powder chamber which unscrewed from the breech, since the state of metallurgy did not permit casting of one piece cannon. However, following some 250 years of use of cast, muzzle-loading cannon, the possibility of breech-loading returned in 1664 when a hinged breech was patented, and in 1704 a French engineer introduced a design that put a hole at the breech, through which the operator could put shot and powder, and which could be closed by screwing in a plug. At roughly the same time an inventor in Milan developed a tip-up breech that could use a paper cartridge, and an American independently invented a similar mechanism.

Although similar developments soon took place in small arms, their effect was not as immediate, because the long range capability of the artillery being developed caused it to have a dominant impact on the battle field. Neither of two developments in small arms, the innovative breech-loading Prussian needle gun or the French Chassepot, which had a range several hundred yards greater, played a decisive role in the Fanco-Prussian War of 1870. (Although –as discussed below – the Prussian needle gun did play a decisive role for Prussia against Austria shortly before that war.)

The decisive factor in the 1870 conflict between the Prussians and the French turned out to be the superiority in use of the Prussian breech-loading rifled artillery over the bronze muzzle-loading cannon used by the French. The Prussians massed their artillery at Sedan, the most decisive battle of the war, and brought the French attacks to a standstill, much as the English longbows had at Crecy. For the most part, French attacks were stopped 2,000 yards short of the Prussian lines, outside of effective rifle range. Once again, range of action had proven the decisive element in war.[17]

The British Establishment Clings to Muzzle-Loading Artillery

The impact of the technologically advanced Prussian artillery was not entirely lost on the British, who set up a special committee in 1870 to study the merits of breech-loading versus muzzle-loading artillery. However, the impact carried only that far, and the committee reported back that it had "no hesitation in giving preference to the muzzle-loading gun."[18]

One of the factors which eventually led to universal adoption of breech-loading artillery pieces was the substitution of slower burning, smokeless powders for the old black powder. Improvements in black powder, to achieve higher power, had led to faster burning, such that maximum pressure was being produced inside the gun before the projectile had begun to move. Guns had to be therefore extremely heavy, or the charge had to be reduced. But slower burning, smokeless powders could be used in larger quantities, leading both to lighter guns and heavier projectiles - if longer barrels could be used to take advantage of the slower burning propellants. Longer barrels however were more inconvenient to muzzle-load.

In the late 1870s, as noted above, British authorities were still clinging to muzzle-loading artillery. The incident that finally forced a change may have been an accident on board the HMS Thunderer in January, 1879. The Thunderer carried two very large guns, a pair of 110- pounders, placed on a 38 ton mount. They were muzzle-loaders, loaded and fired simultaneously. During an exercise, one gun misfired, but this was not noticed because of the nose of the other. Both were then reloaded, and the misfired one double loaded, something that could not have happened had it been a breech-loader. The resulting explosion blew up the 38 ton gun mount and killed 10 people. Two years later, in 1881, the British Admiralty advertised for the first of a series of breech-loading guns, which would be used to combat an increasingly worrisome torpedo boat threat.

Control of Artillery Recoil

An associated development that followed in the late 1800's was a means to successfully overcome the effects of recoil by means of spring and hydraulic systems. The recoil systems not only stored the energy necessary to return the barrel to its original position, permitting a high rate of fire, but soon were redesigned to harness this energy to open the breech and eject the empty cartridge case. With these improvements, artillery, at the turn of the century, again became the dominant weapon. Just as the rifle had earlier gained ascendency over the non-recoiling-absorbing field gun, the gun now temporarily replaced the rifle as, if not a master weapon, at least the dominant weapon of the time. This fact was demonstrated to one eye-witness to the Russo-Japanese war of 1904-1905 who reported to his superiors:

The great impression made on me by all I saw is that artillery is now the decisive arm and that all other arms are auxiliary to it. The importance of artillery cannot be too strongly insisted upon, for other things being equal, the side which has the best artillery will always win.[19]

Delays In Use of New Technology

In spite of the early spectacular successes of artillery, Preston and Wise tell us that development and application of this new technology was less rapid than it could have been. This was due, they say, partly to the high cost of these new weapons, and partly to the conservatism of military men, who found an important source of morale and fighting strength in their pride in traditional weapons. It also appears that success with one weapon, or a particular type or style of one weapon, tended to make a nation slow to see the value of something new or different. For example, as noted above, French muzzle-loading bronze cannon were the scourge of Italy in 1494, but, in later versions, were completely outclassed by Prussian breech-loading guns in 1870.

Gun Powder in Infantry Weapons

The Spanish Matchlock

As another example where possession of one effective weapon apparently made authorities uninterested in a better one, the French, early innovators in artillery, were slow to see a need for hand-held gunpowder weapons. This was left to the Spanish, who developed the arquebus, a handgun with a butt for the shoulder. The arquebus was fired by a match fixed to a trigger. The weight of this weapon was a problem. When early models of the arquebus were constructed light enough for a soldier to handle, they could fire a projectile of less than an ounce, a distance of less than 200 yards. Nevertheless, Spain adopted the arm, and eventually developed an improved arquebus, six feet in length, weighing 15 pounds, which was fired from a fork-shaped rest.

This heavy arm, called a musket, could fire a two and one half ounce ball some 240 yards, but the device had its drawbacks: Not only was it heavy to carry, it required 56 drill movements to reload. After the arquebus (a matchlock) had been fired, the operator had to remove the match from the serpentine, measure out powder and pour it into the barrel, put in bullet and wadding, and push them compactly together. The next steps were to refill the flash pan, close its cover, and blow away any surplus powder. He could then replace the burning match in its holder and fire the gun. [20]

The Spaniards not only adopted and improved the matchlock infantry weapon, they worked out field tactics to use this firearm in field entrenchments, against cavalry and against other infantry. In 1522, at the Battle of Bicocca, the Spaniards deployed their infantry behind a sunken road, and destroyed a Swiss pike force in the employ of the French. In 1525, Spanish infantry maneuvering in the open won a decisive battle at Parma. These events fixed the shape of battle on the continent for the next century: The musket and the pike remained the dominant arms.

> *Artillery opened the way, the pike protected the musket, and the musket cleared the road for the advance of the pike, and frequently also for the cavalry sword and lance. Thus for over a hundred years were tactics set.*[21]

It should be noted that the English were slow to turn to firearms as individual weapons, but more than conservatism was involved. In the early 1500's the longbow was not only less dangerous to its user than the primitive musket or arquebus, but was superior in range, penetrating power, and rate of fire. But finally, 1n 1595, a privy council ordinance declared the longbow was no longer a suitable weapon for the English militia.

A Replacement for the Matchlock: The Wheel Lock

The arquebus with its matchlock was not only difficult to reload and fire, it had additional disadvantages. Rain could put out the match used to fire it, and the glowing match could not be concealed at night. The first answer to these defects was the wheel lock. This had been described in concept by Leonardo da Vinci, but was made operationally useful only in 1517 by a resident of Nurnberg, one Johann Kiefuss. In the wheel lock, a small, spring-driven, toothed wheel turned and struck a spark-producing substance, usually iron pyrites. This relieved the worst problems of the matchlock, and permitted limited use of firearms by mounted troops. But the wheel lock was still not a fully satisfactory firearm. The spring had to be rewound before each shot, and if a misfire occurred, not only did the pan have to be refilled, but the spring rewound again.

The Flintlock

The successor to the wheel lock was the flintlock, a mechanism in which a piece of flint was held in a metal clamp which pivoted to strike the flint against a roughened steel to produce sparks. The sparks then ignited powder in a flash pan. In the early 1600's an unidentified inventor, apparently English, combined the plate and cover of the flash pan so that, when hit by the flint, the hinged cover moved back, allowing the sparks to fall in the powder in the pan.[22]

On one hand, the flintlock is attributed to Simon Marquarte, the son of a German gunsmith who settled in Spain in the late 1500's. On the other, it is often said to have been invented around 1580 by Dutch robbers who found that the glowing match of the matchlocks of the time gave away their night-time positions of ambush. Interesting enough, the early Dutch flintlocks were called "snaphaunce" which is Dutch for "chicken thief." [23]

The flintlock, as often seems to have been the case with something that worked reasonably well, may have caused diminished interest in development of alternatives. At any rate, it continued in use by the British army for some 250 years. The British army retained the flintlock until 1840, passing up an early opportunity to convert to percussion ignition. During this long period of flintlock use, British infantrymen developed faster methods of loading: The learned to dispense with ramming the ball home after pouring powder in the barrel, by simply banging the butt on the ground, which could also force powder through the touchhole and prime the pan.

Percussion Ignition

In the early 1800's various chemists had attempted unsuccessfully to use a class of substances called fulminates in conjunction with chlorate of potash – which had recently been discovered – as a main propellant charge. This mixture's extremely rapid burning, however, made it unsuitable for this purpose. The possibility that the combination of fulminates and chlorates could function, not as a main propelling charge, but as a detonating agent, was not considered until a Scottish clergyman,

Alexander John Forsyth, who was also an avid hunter, encountered difficulty in hunting geese on a lake near his home in Scotland. It seems that, although Mr. Forsyth was a good shot, he often missed because the birds, frightened by the flash of the powder in the pan of his flintlock, flew an instant before the main charge fired. To overcome this problem, he converted his flintlock to percussion ignition, and finding it to be extremely effective, went to London to interest the government in his invention.

His story was well-received by the Master-General of Ordinance, who promised compensation and a place to experiment. However, a government turnover brought in a new group of bureaucrats, and a new head of ordinance, a Lord Chatham, who cut down expenses by sending the troublesome preacher home, without the money promised by his predecessor. As a result of this ill-advised attempt to reduce military spending, the British shelved a significant discovery, and the first military application of percussion ignition awaited the work of a Swiss engineer, the diligence of his German apprentice, and the quick interest of alert Prussian military authorities.[24]

The Prussian Needle Gun

Samuel Jean Pauly, a Swiss engineer living in Paris, experimented with various ways of putting a priming charge and main charge together in one case. A German named Dreyse, who had been apprenticed to Pauly, returned to Germany in 1827 and found the Prussian authorities receptive to the possibilities of this potential breakthrough. Since Dreyse believed that a charge which burned from front to rear would reduce gas leak and impart a higher velocity to the egg-shaped projectile, the detonator was put just behind the bullet, with the powder charge to the rear. The "needle" was a thin firing-pin which, when the gun was fired, penetrated the paper case and the main charge, and exploded the detonator. [25]

Dreyse's original design was for a muzzleloader, but only ten years after his return to Germany, he had put the spring-loaded needle mechanism within the bolt of the first practical breech-loading rifle. Although this Prussian needle gun had a shorter range than its principal continental rival, the French Minie rifle, it had a rate of fire several times as fast. And the operator of the breech-loader could remain in a prone position, without getting to his feet to reload, as the soldiers of other countries, using muzzleloaders, had to do.

The Austro-Prussian War of 1866 demonstrated the superiority of the needle gun over the Austrian muzzle-loading Lorenz rifle. Captured Austrian troops who had to rise up to reload were awed by the Prussian weapon which allowed its gunners to remain hidden while reloading. When the war was over, Prussian authorities were surprised that they had won a stunning victory while using an average of only seven rounds per soldier. Thus came about another example of the effect of a weapon innovation on history.

This conquest of Austria by Prussia increased Prussian manpower by twenty-four million people, resulting in a combined population one third larger than France, which the Prussians would face only four years later in the Franco-Prussian War. [26]

The percussion cap also made possible the expansive metal cartridge case. A Paris gunsmith developed the pinfire case in 1847, only ten years after Dreyse's first breech-loader. The rimfire cartridge was next, and the centerfire case followed in 1861

The Minie Ball

A separate chain of events, this one dealing with projectiles rather than propelling charge, demonstrated the marked indifference exhibited by British government officials of this period to new developments. An Englishman, Captain J. Norton, had observed during his service in India that the natives of South India used lotus pith on the rear of their blowgun darts to give an air tight seal in the barrel of the blowgun, with a resulting higher velocity. [27] He therefore designed a cylindo-conoidal bullet with a hollow base to expand and seal the gun bore. William Greener, an English goldsmith, improved the design by putting a conical wooden plug in the base of the bullet. The British government turned down both ideas.

On the Continent, the idea was taken up by Claude E. Minie, a captain in the French army, who patented the Minie bullet, which expanded when fired, and not only sealed the bore but tightly contracted the rifling in the barrel. This bullet was produced in France in 1849, following which the British paid 20,000 pounds for the right to use the invention.

Rifling, itself made more effective by progress in bullet design, had been invented in

1520, and a gun with rifled barrel produced in Nurnberg. But such firearms were produced only in limited quantities, for sportsmen interested in range and accuracy; they were ignored by military authorities (one reason was cost) for the next 250 years. But then their adoption was rapid. Of these related inventions, Fuller says:

> *These two inventions - the percussion cap and the cylindro-conoidal bullet - revolutionized infantry tactics. The first rendered the musket serviceable in wet weather, reducing misfires in each 1,000 rounds from 411 to 4.5 ... The second caused the rifle to become the most deadly weapon of the century.*[28]

The Development of Advanced Small Arms

Advancements in small arms technology continued in the late 1800's. Development of nitroglycerin-guncotton propellant mixtures permitted smaller bore weapons with higher velocity projectiles. This led to a requirement for a higher rate of spin to stabilize the bullet, but at these higher velocities lead bullets would not pick up the spin of the rifling but simply tore down the bore. The solution to this problem was the copper jacket, invented by E. Rudin, a major in the Swiss army. These developments let to a reduction in the bore of the British army rifle and standardization at 0.303 inches in 1888.

Early Concepts in Repeating Firearms

From the introduction of the first firearm onward, there had been repeated efforts to achieve a higher rate of fire. The most common early approach to the problem was to use a plurality of barrel. For example, as early as the end of the fourteenth century, a multi-barreled anti-cavalry device was built. It combined a number of pike points with twenty or more muzzle-loading barrels, which could be fired simultaneously at charging cavalry. More ingenious were the varied attempts to fire multiple charges from a single barrel. Cleator lists the following examples which show the incessant quest over the centuries for a multiple firing gun:

> *A design of an arquebus by one John The Almain in 1580. The weapon was to "contain ten balls or pellets of lead, all of which shall go off, one after*

another, having once given fire."

A patent issued to Charles Cardiff in 1682 for "an expedient with security to make muskets, carbines, pistols, or any other small fire armes to discharge twice, thrice or more, severall and distincte shots in a singell barrel and locke with once priming, and double locks oftener, reserving one or more shots till occasion offer."

A patent given in 1813 to Joseph C. Champers, of Pennsylvania, for a system of "repeating gunnery" to enable pistols to fire six, and muskets seven shots.

A patent to an Englishman, approximately 1850 for a plan to load a gun barrel with a number of bullets each having a hollow base, filled with powder. The bullets were also drilled through the front end, and a fuse inserted. Igniting the front charge lit the fuse on the second, which fired shortly, which lit the fuse in the next, and so on.[29]

James Puckle's 1718 "Machine Gun"

Eventually, interest began to focus on using a single barrel and putting multiple charges in a separate mechanism. As early as 1718, an Englishman patented a device which involved this principle. James Puckle's machine gun was strangely modern in appearance, with one barrel, supported on a tripod, and a circular plate, turned by a crank, holding several cylindrical chambers, each loaded with powder and ball. In keeping with the spirit of the times, it was designed to shoot round bullets at other Europeans and square ones at Turks.[30]

Samuel Colt's Revolver

In 1818, an American engineer, Elisha Collier, secured an English patent for a complex flintlock pistol which had a spring-driven cylinder, an automatically-replenished flash pan, and used a second spring to move the cylinder back and forth to securely fit against the breech. On a trip to Calcutta, India, to deliver a lecture on laughing gas, another American, Samuel Cold, examined one of the Collier pistols. On the slow voyage back to the United States he designed his first revolver, in which powder and ball were put in from the front of the cylinder's chambers. This innovation was highly successful, but when the Colt patent expired, a competitive firm, Smith and Wesson, introduced a competitive revolver incorporating the improvement of breech-loading.

The Repeating Rifle

The successful repeating rifle was based on a series of American innovations. The first design patent for a repeating rifle went to Walter Hunt in 1849. Ten years later a rim-fire metallic cartridge was developed for use in the Hunt rifle by Benjamin Tyler Henry. The resulting Henry rifle was renamed when the Oliver F. Winchester company purchased the manufacturing rights. The product was a tubular-magazine, lever-action repeating rifle, still of a familiar appearance today.

The box magazine was invented by James Lee and adopted by the British in 1888 at the same time they settled on a standard .303 bore. The combination was called the Lee-Metford rifle after Lee and the developer of the shallow grooved rifling used. When the Metford was changed at the Enfield armory from seven grooves to five, the Lee-Enfield rifle resulted, a firearm which was still in use by British forces in World War II.

The American Civil War gave new impetus to the search for higher rates of fire. The gun named for Richard Gatling of Chicago was originally designed as a rifle with a rotating chamber reminiscent of James Puckle's 1718 device, but it developed into the familiar crank operated array of seven to ten barrels. The cartridges were gravity-fed from a drum container, and each barrel had its own firing and ejector mechanism. The rate of fire was an unbelievable 350 rounds per minute.

Throughout the 1800's, numerous other multi-barrel arrangements were tried, but the revolution that was to come in quick-firing gun technology awaited the visit of an American, Hiram S. Maxim, to Europe in 1881.

The Ultimate Destroyer: Maxim's Machine Gun

At an exhibition in Paris, Maxim ran into another American, now unidentified, who had strong views concerning the propensity of European nations to go to war periodically with one another. He told Maxim: "If you wish to make a pile of money, invent something that will enable these Europeans to cut each other's throats with greater facility." [31] It was sound business advice. Only two years later, Maxim secured a patent for a belt-fed repeating gun whose breech was operated to eject and relock by gas pressure from the proceeding round. Its rate of fire was so high that the barrel had to be cooled by water. Incremental improvement, including the use of smokeless powder and air cooling were added by American designers John Browning, Benjamin Hotchkiss, and I. N. Lewis, and by Baron A. Odkolek Von Augezd of Austria. Cleator sums up the results:

> *In one guise or another, the machine gun quickly showed itself to be the most effective death-dealing instrument yet devised by man, a position it was to hold for a half century or more, during which period, such was the unprecedented slaughter occasioned by its use in the course of two global conflicts, its victims were to be reckoned, not in thousands, or even in hundreds of thousands, but in millions.*[32]

A Near-Master Weapon: Its Impact

The machine gun, when possessed by only one side, gave overwhelming advantage in a conflict. At the Battle of Omdurman in the Sudan in 1898, British soldiers, outnumbered two to one, lost fewer than fifty men while killing 11,000 Dervish soldiers, using 20 Maxim guns and 44 pieces of field artillery.

However, in World War I, when the Western Front had stagnated, the machine gun, possessed by both sides, conferred no such advantage, but simply led to stalemate and wholesale loss of life.

> *Repeated attempts to break through on the part of both sides were defeated with monotonous regularity (and appalling loss of life) by a combination of barbed wire entanglements and interlaced machine gun fire, this last the unforeseen cause of nine out of ten casualties suffered by the contestants. Nor did efforts to obliterate these obstacles by concentrated artillery shelling solve the problem, as the intensive bombardment of a particular sector of the line served to give warning of an impending infantry attack. Invariably this would be repulsed by defenders*

who had taken cover in an even stronger position in the rear, and for the next three years the combined backward and forward movements of the opposing forces, undertaken at the cost of hundreds of thousands of lives, covered a distance of less than ten miles.[33]

French Obsession with the Offensive

Blind exaltation of the offensive started with the doctrines of a Colonel Ardant du Picq whose military works gained prominence in France at the turn of the century. Essentially, he argued, the issue did not depend upon weapons so much as upon the attitudes of the contestants. At the beginning of the twentieth century military thought in France was largely guided by Ferdinand Foch, than a lieutenant-colonel at the Ecole de Guerre (Military School) and acquiring a rising reputation as a military theorist. Foch picked up these conclusions of du Picq, including a blind acceptance of the superiority of the offensive in all its forms, tactical as well as strategic.

In time Foch became the protagonist of the frontal offensive delivered with maximum violence and at all cost – the *attaque a outrance.* By the time he was given high command in the field he had already formed his military creed – Attack! Attack! This was his solution to any given military confrontation. The influence of this teaching at the Ecole de Guerre was strongly felt by the progressive elements among the French officers of those days, the "Young Turks" whose leader Colonel Grandmaison was a favored pupil of Foch.

It was Grandmaison who, as Chief of Operations in 1913, was mainly responsible for the adoption of the incredible French plan of campaign in 1914 – Plan XVII, which envisaged a headlong attack eastward of the whole French field army as a counter to a massive German invasion.

A combination of these influences served to bring about a decline in strategic thought and resulted in suicidal and purposeless offensives in a purposeless war which, once started, no one seemed to have either power or the will to stop.

The French, under Foch's strategic guidance, fought even their defensive battles by a series of violent offensives, as at Verdun, losing considerably more casualties than the Germans. In the great offensives of 1915, which Joffre delivered ceaselessly against the unyielding Germans, the French casualties mounted to over a million and a quarter while inflicting upon the Germans only about one third that number. It was the same during most of 1916 and 1917 – at the Somme, at Arras, at Passchendaele – till complete demoralization set in and ended with open mutiny in sixteen army corps of the French army.[34]

After that horrific conflict, it is not difficult to understand that strategies such as those of Doughet and Mitchell – emphasizing the importance of air power – might arise from a search for alternative ways to win a war. Even easier to understand is the reaction of the French, who had entered World War I with an irrational but firmly-held belief in the power of "elan", the mystique of the charge, and who would switch after that war to an equally strong fixation with defensive fortifications, resulting in the Maginot line.

It needs no argument to prove that, should other things be equal, a man armed with a sword alone is no match for an opponent armed with sword and shield. This simply means that protective offensive power is superior to unprotective offensive power.

J.F. C. Fuller

CHAPTER 6

THE SEARCH FOR A BETTER SHIELD

Metal Armor and Arms

Protective armor of animal skins has been used since the earliest times. Eventually, when the difficulties involved in making and shaping large pieces of metal could be overcome, metal armor began to come into use. Due to its relatively low melting point (1,083º C) copper found early use by primitive people and was later supplanted by bronze. Iron was a much later development. In fact, it may be that the use of copper with tin to form bronze was one of those instances where a particular technology turned out to be so successful that it delayed the introduction of a more advanced technology, in this case iron.

Primitive Metallurgy

The names given to iron by primitive peoples reveal the source from which they obtained it. The Summerians named iron the "metal from heaven" and the Egyptians called it "black copper from heaven.". The limited amount of iron used by these societies came from meteorites, most of which contain iron in association with other substances, particularly nickel. [1]

Man-made iron, on the other hand, seems to have first occurred as a fragmentary, dross-like by-product of pottery manufacturing, which at first was not recognized as being the same substance as the "metal from heaven." Because of its relatively high melting point (1,535º C) iron could not be obtained by primitive people in a molten state. Neither the Egyptians, Mesopotamians, Greeks, or Romans managed to melt iron, however it was discovered that hematite-native anhydrous ferric oxide, Fe_2O_3, an important iron ore, could be reduced at the highest heat then obtainable (around 1100º C) to a mixture of iron globules, cinders, and slag. Constant reheating and hammering purified the impure metal somewhat, but the end product was still a relatively soft, slag-contaminated material much less hard than bronze. For centuries iron remained, not a utilitarian metal, but a rare and expensive ornamental metal, at one time valued at forty times the price of silver and five times that of gold. [2]

The meeting of two dissimilar levels of technology is illustrated by the conflict between the Celts and Romans, long after the Assyrians, Greeks, and Romans had drastically improved iron processing and the Celts were still at a primitive stage in iron-working technology. It is said that the long Celtic swords bent so easily that their use was effectively limited to a single downward stroke. Then, if he had time, the user was forced to straighten the blade by placing between his foot and ground, after which, if still able, he could deliver a second blow. During that period of Roman weapon superiority, there was little doubt of the outcome in most battles between warriors of the two societies.

Steel

Even without the facilities to create molten iron however, primitive peoples did make a series of discoveries that led to a superior form of the metal. First, they discovered that iron undergoing sustained heating in the presence of carbon acquired a hardness it did not have before. Then, the opposite of what would be expected from experience with the more familiar bronze or copper, the iron item was found to become harder still if quenched by plunging it into cold water. When the Romans discovered that overdoing this process could make the material brittle, they developed the process of tempering, making the quenching less severe or using careful and moderate reheating.

A considerable amount of mysticism accompanied the search for exotic quenching fluids, leading at times to almost unbelievable brutality in slave societies. In others, the result was only odd, as in the case of the old Sheffield firm of steel makers who held that camel urine was unsurpassed for this purpose, such that it was imported by the barrel load from Egypt.[3]

Once developed, hardened forms of iron gave a pronounced advantage to an army in battle. The use of the conventional bronze thrusting sword, for example, was governed by the strength of its blade, but the force of the hardened iron sword's slashing stroke was limited only be the strength of the human arm which wielded it. In fact, it was discovered that a weapon made of the new material could cut right trough a sword made of bronze.[4]

Shield and Body Armor

The Greeks, even before the Romans, brought shield and armor to an advanced state. The equipment of the Spartan soldier – including helmet, cuirass, greaves, and shield – weighed some seventy two pounds. Nevertheless, although moderately protected from the sword blows of opposing infantry men, the Greek soldier remained vulnerable to skirmishers armed with javelins who would not stand and fight ("stay and be slaughtered" as Cleator puts it) but who chose to retreat and throw their javelins from a distance.

This vulnerability foreshadowed that of the army of Harold of England at the battle of Hastings in 1066, described earlier. As the reader may recall, Harold's army, bearing a hedge of shields and wielding battle axes could be charged by William's cavalry only after it was decimated by high trajectory arrows fired by the archers in the Norman army.

Because the individual Greek soldier depended for protection on the shield held by his companion on the right, there was a tendency to edge to the right, with the result that the entire phalanx tended to move to the right once the enemy was engaged, leaving an exposed left wing. Cleator describes the part that this characteristic played in the battle of Crnoscephalae, between armored Greek and Roman infantry in 197 B.C. There the Greek army of Philip V of Macedon met an essentially equal force of Romans under the Roman General Quinctive Flaminius. Neither army could successfully assault the other from the front, and the battle was without resolution until "an unknown tribune detached twenty maniples from the Roman right wing" – which had apparently edged completely off the battle front – "led them behind the enemy lines, and charged from the rear." [5]

Although Greek and Roman tactical formations were different – the original Roman copy of the Greek phalanx having given way around 300 B.B. to the more flexible maniple – Roman armor was similar to Greek. The legionaries' armor included a metal helmet crowned with feathers and a bronze breast plate. Their oval shield was made of wood, covered with hide, and strengthened with a central boss and rim of iron.

Armor in the Middle Ages

In even a brief discussion of armor, mention should be made of the attention given to it in the Middle Ages by Charles the Great of France. Charles, after he assumed the throne in 768 A.D. conquered and held a vast empire that stretched from the Elbe River to the Pyrenees and from the English Channel to south of Rome. He used many innovative approaches to improving the equipment of his forces. He divided his kingdom into districts, each picketed by a number of fortified posts, drastically improved the weapons with which his horsemen were armed, and even gave attention to better weapons for his foot soldiers. Where formerly medieval "infantry" had been merely a rabble carrying clubs and agricultural implements, that of Charles was armed with sword, spear, and bow. Charles also decided that an army could not be adequately mobile and fully effective if it (1) had to rely on foraging, and (2) could not storm walled cities when necessary. He therefore put together a supply train carrying rations for three months and clothing for six, and a siege train for use against fortified cities.

But it was to body armor that Charles gave his greatest attention. He set up a census of the armor in the kingdom so that none could be hoarded and also made laws to prohibit its export.

The Developing Technology of Medieval Body Armor

One early solution to the problem of making and shaping the large pieces of metal body armor was the use of overlapping plates of metal attached to a strong and heavy undergarment. A considerable lighter form of armor arrived on the scene with the development of protective garments made of iron rings welded or riveted together. In the East this development was further improved by interlacing the links, each of which was passed through four others, doing away with the need for a support garment. With finely wrought rings, the armorer could not only make a protective suit with no openings, but one that could be shaped to cover any part of the body.

This chain mail was particularly effective when worn with a quilted garment beneath, and was the main defense of the mounted warrior from the eleventh through the thirteenth centuries. The principal vulnerability to the heavy sword wielded by opposing knights of the time was bruising or even broken bones if the assault were forceful enough, sometimes accompanied by the driving of broken links into the wound.[6]

Chain mail was supplemented by the use of metal helmets and improved by extra protection of the joints by molded leather. Eventually however chain mail was discarded in favor of greatly improved plate armor. Armorers, as they developed better tools and material to work with, began to produce complete suits of articulated protective plate armor, a form which was carried to a high degree of development in the 15th and 16th centuries. From the beginning of the 15th century on, barons and knights typically wore full suits of armor plate. By the year 1500 the product had reached a peak of development and particular localities such as Milan and Nurnberg were famous for their armor. King Henry VIII, who ruled England from 1509 to 1547, set out to break the continental monopoly on high quality armor by importing armorers from both Italy and Germany and setting up his own workshops at Greenwich. The resulting "Greenwich" armor was of very high quality and easily recognizable by its "humpy" appearance at the joints.

The armored knight became nearly invulnerable to the weapons permitted by the customs of the times, as long as he was on his horse. A notable disadvantage – in addition to cost – was its weight.

> *The wearer was by this time so heavily encumbered (with a load of anything up to 100 pounds) that he could scarcely raise his weapons, and death from heart failure became an additional hazard of war for the elderly, while any knight who had the misfortune to be unhorsed could but lie where he fell, an easy prey for the first predatory foot soldier who came*

along.[7]

Fortifications

It is not possible to provide here a comprehensive discussion of the subject of fortifications. Only a few examples will be treated. It is worthwhile to note however that an identical outside threat can lead to different defensive responses, depending on the characteristics of the society being threatened. For example, although both devoted increased attention to defensive fortifications, England and France otherwise responded quite differently to the Viking incursion that began in the late 700's.

The Viking Threat to Europe

The first scattered Viking raids took place soon after Charles the Great came to the throne in France, but it was not until after his death in 814 that the problem became severe. In 850 A.D. "a storm hit the European continent":

> *In 850 the whole manhood of Scandinavia took to the sea, and the half century which followed was one of the darkest periods in European history.[8]*

In response to this threat, castles and strongholds were built on the continent, towns were walled, and bridges across rivers were fortified. Since only mounted men could move rapidly to meet to meet the raiders, the mounted knight assumed increased importance.

The English also built fortifications in response to the Viking threat, but there the similarity to the French reaction ended. King Alfred, who ruled for the last half of the 9th century (848-900) built a fleet and defeated the Vikings in their chosen element. As a result, while cavalry was developing on the continent, the English, for land conflicts, continued to rely on infantry, a difference which persisted, and was demonstrated in the Battle of Hastings in 1066. William the Conqueror, we should note, was the three times great grandson of Rolf, one of the fiercest of the Viking chieftains, who plundered, conquered, and eventually settled in Normandy during the period of Viking invasion.

Development of the Castle

In its original form, the European castle, the fortified residence of an individual, as a thing apart from a walled city, was little more than a moat-encircled timber keep, built on top of an artificial mound of earth. Gradually, stone towers came into use, and later concentric complexes of stone walls overlooked by stone towers which were first square and later round. With the use of stone instead of timber, and increased size and complexity, the cost of these structures increased accordingly. Although seemingly invincible, such fortifications could be defeated by an extended campaign, itself quite costly, in which the attacker, if he could prevent both breakout and relief, could eventually starve the defenders out.

Typically, the attacker would build a wall around the castle to prevent the occupants from leading an army out and another around his own camp for protection. Such actions were customarily accompanied by a propaganda campaign with offers of generous treatment should be the gates be opened and warning of punishment if they were not. Perhaps there is an inherent desire in human beings to believe the pleasant lie, rather than undertake the rigors of a sustained defense.

> *Surprisingly, in view of the fact that past history provided a sorry tale of treacherous disregard for the most solemn of undertakings given to an adversary at bay, such a resort to psychological warfare was often effective.*[9]

If starvation, psychological warfare, or treachery did not bring about success, the attackers could, if the objective were sufficiently important, resort to mining, bombardment, battering rams, and scaling implements. But the expense, time, and manpower required generally made the castle an effective defensive fortification.

Inherent Strength In a Strong Defensive Posture

Palit's analysis of Clauswitz's conclusions on defense can be summarized as follows: A defensive attitude is not necessarily always preferable to an offensive attitude, but the defensive is the stronger form. Therefore, although the defensive cannot itself achieve a decision, it should be used by the weaker side to restore the balance, and prepare for an eventual campaign to enforce a decision.[10] If this is correct, then the continued reappearance of defensive constructions should be expected through the ages. And such seems to have occurred.

For example, General Lee's forces, on the defensive in wooded country between Washington and Richmond, adopted the practice of setting up positions behind high wooden breastworks. In this position, though Union scouts might infiltrate, formation assault became too dangerous to carry out against accurate small arms fire from the defenders. Palit believes that this tactic, made particularly effective by improvements in small caliber firearms, foreshadowed the defensive war that would take place in Europe a little more than half century later.

> *Assaulting infantry could no longer hope to take defensive positions with the bayonet. The old shock tactics continued to be tried out by die-hard generals as, for instance, at Gettysburg or in the wilderness campaign – but assaulting formations were always broken up or were wiped out by withering fire from the breastworks ... The days of shock action were over – or should have been.[11]*

By the close of the 19th century the tactical balance had swung even further in favor of the defensive, aided chiefly by the introduction of the machine gun. In World War I interlocking fire from this weapon, laying down a literal curtain of bullets in the path of assaulting troops, often held up by intricate lines of barbed wire, made infantry assaults on strong defensive positions all but impossible.

> *But this basic tactical truth was hidden from most military men by a myth – the myth of "cold steel". There existed among the military castes of Europe a blimpish belief in the heroics associated with the bayonet charge in massed assault formation ... Throughout the first world war, commanders who in many cases never went near the front lines but issued their orders from headquarters situated many miles to the rear, kept on sending their men "over the top" in close formation, to charge the enemy's defenses in frontal assault.[12]*

Spear As Shield

Typical illustrations of the Macedonian phalanx on the march give the formation much of the appearance of a porcupine with quills pointed to the front, with long spears taking the place of quills. Philip of Macedonia doubled the depth of the traditional Greek phalanx, creating the grand phalanx, and gave his infantry a much longer pike or spear. This, depending on the historical authority, is described as ranging from twelve to twenty four feet in length.

According to Creasy, this "Sarissa" as it was called reached out eighteen feet in front of the soldier who carried it. If ranks were two feet apart, the spearheads of six rows of soldiers extended beyond the front line, "presenting a bristling mass of points" to the enemy.[13]

Even the shorter, traditional Greek spear had been a highly effective defensive weapon – not of course as a defense against projectiles but as a defense against direct assault by other warriors. This was demonstrated at the Battle of Marathon, where the spear was wielded by the well disciplined Greek hoplites, who went into action on the run and closed with the Persian army before the Asiatic bowmen of King Darius could organize their fire. Once at close quarters, the Greek spear's effectiveness as a defensive as well as an offensive tool is illustrated by Creasy's description of that battle.

> *While their rear ranks poured an incessant shower of arrows over the heads of their comrades, the foremost Persians kept rushing forward, sometimes singly, sometimes in desperate groups of twelve or ten upon the projecting spears of the Greeks, striving to force a lane into the phalanx, and to bring their scimitars and daggers into play. But the Greeks felt their superiority, and though the fatigue of long-continued action told heavily on their inferior numbers, the sight of the carnage that they dealt amongst their assailants nerved them to fight still more fiercely on.[14]*

Stakes and Spears

Charles the Great of France, mentioned earlier in connection with the development of body armor, was responsible for yet another innovation in medieval defensive warfare. In an age when the chief threat to his siege and supply trains was cavalry, he equipped these trains with bundles of iron-shod stakes as a protection against mounted attacks.

Chapter Three briefly discussed the unexpected defeat of heavy cavalry when confronted by archers skilled in the use of the longbow. Developments in use of the pike, or spear, did apparently hasten the decline of the mounted knight. The "despised peasantry" when pressed by continental nobility had, on occasion, used improvised weapons similar to the farm implements with which they were familiar. From these beginnings and from the more traditional spear and battle axe, a group of pole arms developed, shaped to thrust, hook, or chop at attacking horsemen. One historical illustration shows fifteen different pole arms featuring points, hooks, and cutting edges, all at the end of long poles.

The Swiss furnished the first recorded example of large scale use of the pole arm in battle when they staged an uprising that threw off the rule of King Leopold I of Austria in the early 1300's. When the Hapsburg monarch sent his powerful cavalry against the Swiss rebels in 1325 the mounted troops were met by commoners with little knowledge of horse and armor, who were outnumbered ten to one. It should have been no contest. But the Swiss footmen attacked the invading cavalry column, pulled the knights from their horses and hacked them to pieces, and drove their infantry into the waters of adjacent Lake Egeri. [15]

Other successes followed, and for decades the Swiss were highly regarded in the field. The legend of Swiss invincibility which was built up by Swiss pikemen lasted until 1515 and the Battle of Marignaro. There, the French conclusively demonstrated the vulnerability of massed pikemen to a new tactic using a new weapon, a relentless barrage from primitive French cannon.

Although the pike passed from the scene as a central weapon of war because of the vulnerability of the pikemen to fire from gunpowder weapons, it remained in altered form as the bayonet, a device named after Bayone in Southern France.

When the longbow was reluctantly abandoned by the English in 1595 in favor of the musket, its accompanying pikemen remained in the British army until 1702 when the bayonet was adopted. This weapon enabled two types of infantrymen to be combined into one and provided the musket-armed infantryman with his own spear-point protection. It seems to have originated more or less by accident on the continent when some unidentified foot soldier stuck the handle of his dagger in the barrel of his flintlock to defend himself from an attacker who arrived before he could reload. This early plug bayonet, with its obvious disadvantage, was soon replaced by the socket bayonet, which left the muzzle open to reload and fire.

Limitation of Spear as Shield

The Greek phalanx, as described above, was virtually unassailable as long as it retained an unbroken front. On rough ground however, where gaps in the line were likely to appear, opportunities could exist to breach the hedge of spear points and the line of shields. Once that was accomplished, the length of the Greek pike made it nearly useless. An attacker, armed for example with the two foot Roman sword with its obtuse point used mainly for stabbing, could do extensive damage once inside the spearman's guard.

The Importance of the Shield

Cleator describes a brutal experiment that demonstrated for all time the change in balance of power that occurs if two adversaries, one armed with spear (long reach weapon) and one with sword (short range weapon) are each given a shield. In the first case, where neither has a shield, the spear can act not only as long range weapon but as protective device to keep the opponent beyond the lethal range of his short range weapon. But if the long range weapon can be made ineffective, the tables are turned:

> *Whatever the outcome of a duel between antagonists armed with sword and spear alone, when a shield is added to their equipment, the advantage unquestionably lies with the swordsman, as the longer reach of the spear is effectively neutralized. This was demonstrated to his entire satisfaction little more than a century ago by the Zulu Chief Chaka. He armed a hundred of his followers with shield and short thrusting assegai, and another hundred with shield and long assegai (throwing spear) and set them to fighting one another. Those given the short thrusting assegai (the native equivalent of the sword) secured an overwhelming victory over their opponents, and from then on Zulu warriors were armed accordingly.[16]*

The Maginot Line

In even a brief discussion of shields and defensive fortifications, some attention must be given to the much derided defense against land attack which was devised by the French after World War I – the Maginot Line. As noted earlier, the French had entered World War I with a blind faith in "elan", the spirit of the offense [17] and as a result saw their army bled white in the barbed wire and machine gun-dominated trench warfare of that conflict.

It was natural swing of the pendulum after that war for the French to prepare a system of super trenches, a network of interconnected underground fortresses along France's eastern frontier, intended to stop any German attack on their border. Perhaps some of the derision heaped upon the designers of that defensive wonder is deserved for two reasons: First, for not finishing the line, and second, for thinking that any weapon or device is necessarily the last word in warfare, since history seems to show that unwavering faith in any weapon or tactic will in time be turned into a false hope by new technologies, countermeasures, or changing economic conditions.

But in the case of the Maginot line, as it was built, it was not necessary for the march of time to bring obsolescence to pass. The line was simply inadequate the day its builders pronounced it done, due to their unwillingness to build a complete system. To be effective, the line would have necessarily stretched from the Mediterranean Sea to the North Sea. [18] In fact, it was built only from Mulhouse near the Swiss border to Montmedy on the Belgian border. To justify not continuing the line along the Belgian border, it was argued that the lowly Ardennes would be a sufficient barrier to invasion. In June 1940 the German army simply encircled the line by way of Belgium and took it from the rear in a matter of weeks. So much for "strategic sufficiency".

Introduction of the stirrup from the Mediterranean world sometime in the seventh century greatly increased cavalry efficiency, since it enabled a firmly seated rider to deliver a powerful thrust with his lance or to rise in his saddle to use his sword with greater leverage. The infantry of the Macedonian phalanx or the Roman legion had never had to face such firmly seated horsemen ... Not until the time of the Swiss halberdier of the fourteenth century could infantry again cope successfully with cavalry on its own terms.

Robert Preston and Sidney Wise

CHAPTER 7

THE QUEST FOR MOBILITY

Early Attempts to Achieve Mobility

Along with the quest for range and the search for defensive protection came the desire for speed of movement. Mobility of men and weapons appears to have been the third consistent thread in the development of tools of war. An example is furnished by Roman improvements to their infantry forces to improve flexibility and movement on the battle field. The original Roman legion, or 'gathering of the clans' was patterned after the old Doric phalanx. It consisted of some 4,000 men in eight ranks, flanked by two small groups of cavalry of 150 men each. As in the case of the Greek phalanx, for its effect it relied on shock in the frontal engagement of two armies.

A large, solid block of men, it moved unevenly over rough ground, and any sustained pursuit of an enemy was difficult. References indicate that this formation was changed during the Gallic wars of 391- 360 B.C. when the Romans were faced with two problems: rough terrain and the Celts themselves, described variously as either "the Celtic sword phalanx" or "hoards of undisciplined tribesmen." [1] In response to this threat, the legion was divided into three groups in the order of battle. The front-most formation was made up of groups of young soldiers with some combat experience, the next group was made up of seasoned veterans, and the third of a mixture of older men and novices. Separate blocks or "maniples" were arranged into ten ranks and twelve files of soldiers. Gaps were left to the right and left of each block. The maniples of veterans followed the groups of younger soldiers, in the gaps between them, so that, if the first formation was driven back, it could retire between the maniples of veterans, and so on. The mobility achieved by this formation change was used to rapidly alter the position of soldiers, to increase their total effectiveness.

A second type of mobility involved the ability to move weight or bulk, rather than people, rapidly across the battlefield in order to overcome infantry flesh and bone with the shock of metal, or with the shock of heavier bodies such as those of horses or elephants. Although Alexander the Great faced armies using elephants, he did not choose to employ them with his own forces. However, one of his successors, Seleucus, valued them so highly that he traded one of the eastern provinces conquered by Alexander for a herd of 500 elephants and used them successfully. Elephants seem to have been a weapon for which effective countermeasures were possible and which were most effective when used as a surprise weapon for the first time against an enemy which did not expect them.

For example, antagonists developed instruments to injure the elephants' feet. The award for bizarre and cruel creativity in this regard must go to the Megareans however: "They drove against Antigonus's elephants pigs smeared with pitch and set alight." [2] As a counter-counter measure, that ruler ordered his East Indian Maouts to always keep swine with the elephants to make them more familiar to the giant beasts.

Antiochus I apparently used elephants as a surprise weapon against the Gauls ("I am ashamed to think that we owe our safety to these sixteen animals")[3] The elephant however, could do great damage to its user's cause if it became uncontrollable. An account of their use by Antiochus III indicates that although that monarch used a force of Indian elephants to overcome a force of African elephants fielded by Ptolemy IV in 217 B.C. , in a subsequent battle some 27 years later his elephants became unmanageable and threw his own army into confusion, leading to his defeat. A similar problem is believed to have afflicted Hannibal at the Battle of Zama in Africa in 202 B.C. during the second Punic War.[4]

The Horse

The "wild ass of the mountains" introduced into the Near East by nomadic people from the northern steppes proved to be a more dependable weapon of war than the elephant. It is believed to have been used by the Kassite people of Babylonia, and its domestication apparently spread outward from that center.

Accounts of the history of the horse indicate that its use as a chariot-puller probably preceded employment as a cavalry mount. Here also, tactical surprise may have been a factor in its success. The Hyksos capture of the Nile Delta may have depended on the use of war chariots against an enemy with no experience in combating them.[5] By contrast, centuries later, although Darius counted heavily on his scythe chariots to disrupt the disciplined infantry fielded by Alexander the Great at Arbela in 331 B.C. this did not occur. Alexander had trained special troops to counter the weapon.

At the opening of the battle, when the chariots were sent clanking across the plain which Darius had forced his followers to level by hand for the occasion, special, lightly-armed Greek troops went into action. They wounded horses and drivers with missile weapons and then ran alongside cutting the traces and seizing the reins to disrupt the change.[6]

Other references to the early use of chariots are found, including their employment by ancient Egypt's first standing army and by other armies of the Middle East,[7] and the alarm created among Caesar's legions by war chariots of the Celts.[8]

The Horse As Cavalry Mount

The use of mounted soldiers is found at least as far back as Assyrian times (illustrations have been found of use during that period) but Cyrus the Great was the first to organize a group of riders for the express purpose of pursuing and overriding a retreating army driven from the field by his archers. This tactic also proved popular with the Persian rulers who followed. For example, Xerxes, son of Darius the Great, is said to have fielded 80,000 cavalry riders against the Greeks.[9]

The names of these Persian rulers can be very confusing. Here are the ones most used in this book: Cyrus the Great, 600-529 B.C. founder of the Persian Empire. Darius the Great, 559-486 B.C. who was king of Persia from 521-486 B.C., and Xerxes, son of Darius the Great, 519-465 B. C., king from 486-465 B.C. Xerxes I is most likely the Persian king identified as Ahasuerus in the biblical Book of Esther. As noted in passing earlier, he invaded Greece in 480 BC. Like his predecessor Darius I, he ruled the empire at its maximum extent, even briefly conquering more land on the Greek mainland, gains then given up through losses at Salamis and Plataea.

> *The irresistible Macedonian Horse supplied the margin of victory and made Phillip master of Greece. This army, which Alexander inherited two years later in 336 B.B. , was vastly superior to that of the Persians which it would face. The element in the Macedonian technique of conquest ... which constituted the most significant advance on traditional Greek warfare, was the combination of the rocklike phalanx with light and heavy cavalry. The union of an always dependable infantry base with the mobile shock supplied by cavalry was too much for the valorous but heterogeneous Persian masses, whose chief advantage, numbers, was greatly outweighed by steadiness, missile fire, cavalry shock, and generalship.*[10]

Centuries later the efficient Eastern Romans would combine the mobility of the mounted soldier with the most powerful long range weapon of their day to create the Byzantine horse archer, discussed earlier.

The bringing together of the equipment developments that eventually made the horse so powerful on the continent of Europe was a slow process. Reins, spur, and bit appear to have been in use at the time the Assyrians mounted part of their infantry. But iron horseshoes, used by the Celts, were not adopted by the Romans until shortly before the time of Christ. The Romans did not substitute saddle for horsecloth until the 4th century A.D. and did not adopt the metal stirrup until approximately 500 A. D.[11]

The Horse as a Master Weapon

Once established as the primary force in European warfare however the horse tended to remain the focus of tactics. Two reasons appear most important: (1) European war by the Middle Ages had become chiefly a monopoly of the nobility and only they could afford horses, and (2) sensibly enough, this group became quite defensive-minded, in a personal sense, and the horse permitted the individual warrior to carry a considerable amount of armor. Although there was evidence that the horse and mounted knight combination did not constitute a master weapon, its status as such persisted - perhaps facilitated by the fact that armored European fought armored European, both using the horse - until the English introduced the longbow. Even the astonishing success of Richard Coeur-de-Lion at Acre with a mixed force did little to disturb traditional European thought.

In that battle in 1192, during the Crusades, Richard, with only 2,000 foot troops and 55 knights, received word that he would soon be attacked by 7,000 Mamelukes. (The Mamelukes were members of a feared military force, originally made up of slaves, that seized power in Egypt about 1250, ruled there until 1517, and remained powerful until the early 1800's.) He deployed his meager force to meet the attack by putting a line of kneeling infantry spearmen in front, and immediately behind them, with room to fire between the spearmen, a line of cross-bowmen, and behind them, a line of cross bow loaders to maintain the rate of fire. The lines of pikemen and bowmen kept the attackers at bay, killing 700 of them along with 1,500 mounts. When the attacking force broke and fled, Richard and some fifteen knights charged and chopped their way through the mass of confused horsemen and back again. In this battle, which apparently escaped the attention of military leaders on the continent, Richard lost two men. In spite of incidents such as these, the mounted combatant, once established in Europe as a master weapon, continued its dominance.

> *From the time of Charlemagne onwards, the plains of Europe ... were dominated by armoured cavalrymen whose shock tactics no foot soldiers of the day could withstand, and though infantry continued to play a part, ... their usefulness was limited except during siege operations ... and mountain warfare. On the field of battle, mounted rider faced mounted rider, each armed with lance, sword, and mace.[12]*

Ships Powered by Oar and Sail

The search for mobility was demonstrated in naval warfare also. Pottery attributed to forerunners of the ancient Egyptians, dating from 3500 B.C. contains pictures of boats rowed by many oars. The first use of sail is also attributed to these people, the invention being stimulated by the fact that the River Nile flows generally opposite to the prevailing wind. Trips could be made upstream propelled by the wind and downstream with the current. From the region of the Nile, the building and use of ships is believed to have spread throughout the Mediterranean.[13] The Phoenicians were early and persistent sailors of the Mediterranean and the Greeks by the 8th Century had built a class of warships driven by 50 oars, called the Pentecontor. This was followed by the bireme, propelled by two banks of oars, and, in 550 B.C., by the Trireme, about 140 feet long, driven by three banks of oars and fitted both with bow ram and on-board space for a contingent of marines. The Athenian fleet played an important part in resistance to invasion by the Persians and also in the commercial development of Athens.

Carthage, a later Mediterranean power, was a trading nation of expert sailors and competition with that power forced the infantry-minded Romans into an attempt to match the Carthaginians in their own environment. The brute force approach adopted by the Romans was costly but effective. They made copies of a Carthaginian trireme that had been driven ashore and learned to operate these copies by trial and error. Inexperience and resulting mishandling of their ships in bad weather led to costly losses, but they made up these losses by drawing on the extensive material and human resources of the Italian peninsula. They then used grappling hooks and swinging gangplanks to turn their sea battles into boarding encounters. With this approach, Rome became a naval power, an advantage that later became critical when Scipio Africanus used Roman control of the seas to attack Carthage itself and force Hannibal to give up his campaign in Italy and return home.

As discussed briefly in Chapter Four, the Eastern Roman Empire continued as a great trading nation long after the demise of Rome and the Western Roman Empire. The Byzantines found a powerful navy to be a necessity in protecting trade routes and ports.

The use of naval power by Phoenicians, Persians, Greeks, Romans, Vikings, and others could fill volumes. However, in the next chapter, we will move forward to the 15th and 16th centuries when England proved the importance of sea power to that island nation. One classic battle will illustrate the advantage that mobility and quickness can bestow on the antagonist who possesses that attribute.

Chapter 7 – The Quest For Mobility

The rise of England as a naval power followed the invention of gunpowder, which was discussed in terms of land warfare in Chapter Five. Fuller believes that the advent of gunpowder, and the weapons to use it, was a primary cause of the rise of the nation state, and the beginning of mass warfare in the modern world. The high cost of equipping armies with artillery and arquebusses was too much for the splintered elements of feudalized societies to bear. Since England, in the War of the Roses, destroyed the remnants of feudalism, while that institution still persisted on the continent, it was in a position to take the lead in this "armament revolution." [14] Influenced by its island location, England began to develop as a sea power.

In sheer numbers the Spanish Armada that sailed to conquer England dwarfed the tiny British fleet. The story of its defeat illustrates the critical importance not only of superior technology but of brave combatants and intelligent leadership. It also reminds us that miserly behavior by a head of state has the potential to trump all three.

Desperate affairs require desperate measures.

Horatio Nelson

Whoever can surprise well must conquer.

John Paul Jones

CHAPTER 8

SAIL AND GUNPOWDER

English naval ascendency had its roots in the interest taken by Henry VII (1485-1509) in building new ships, and the realization on the part of his son, Henry VIII (1509-1547) that England's northern seas were more suited to sails and cannon than to oars and boarding tactics. Henry's confiscation of church property provided funds for ship-building and equipage. Gun foundries were established and put on a permanent footing to provide guns for his great ships. Two examples were the *Great Harry* and the *Henry Grace a' Dieu*. The *Great Harry* carried four 60 pound cannon, a number of 35 pounders, and numerous smaller guns. These fighting ships marked a new concept in naval thought, ships that would fight at a distance as opposed to the conventional of warships as "floating forts" carrying garrisons of soldiers to board and capture enemy vessels. [1]

The technological advancements set in motion by these two English kings provided the environment for development of a still more advanced type of ship that would later protect England from invasion by a numerically superior Spanish fleet – in spite of the indecision and shortsightedness of their monarch, Queen Elizabeth.

The Spanish Monarch Determines to Invade England

Spain in the days of Phillip II (1556-1598) was the leading military power of the

Western World. It possessed a powerful standing army, the only one in Europe, and the world's largest fleet, built to protect the shipments of precious metals that poured into the Spanish treasury from Spanish outposts in Peru, Mexico, New Spain, and Chile. In 1850 Phillip captured Portugal, adding its rich maritime empire to his own. The one flaw in Phillip's world was England. English privateers raided his colonies, defeated his fighting ships on the high seas, and even burned Spanish arsenals on the coast of the homeland itself.

And England was a thorn in the flesh for yet another reason. Phillip was a fervent Catholic at a time when Protestantism was in retreat on the continent and England alone remained as the center of Protestant power. The Pope, Sixtus V, urged the Spanish on, and ultimately established a monetary reward to be paid when Spanish troops would step onto English soil to depose the heretic Elizabeth.

With the execution of Mary, Queen of Scots, in England, an attempt at Spanish invasion of England became inevitable. [2] A Spanish invasion fleet was built and, although transparent attempts were made to disguise its ultimate objective (it was announced at one time that it would go to the Indies, and, at another, that it would sail against English-supported rebels in the low countries) the one purpose of the Armada was, from the first, to support and supplement an invasion of England by a Spanish army gathered on the coast of Flanders by Spain's leading general, the Price of Parma.

Royal Rivalries

When we consider the wealth and treasure that went into building of the Spanish Armada, it may be worthwhile, as an aside, to take a brief look at the intense rivalries in the royal families of Spain and England that fueled Phillip's desire to conquer England.

Mary I was the only child of Henry VIII and his first wife Catherine of Aragon who survived to adulthood. A Catholic, she was Queen of England and Ireland from July 1553 until her death. Because of her executions of Protestants she became known as "Bloody Mary". She was preceded on the throne by her younger half-brother, Edward VI, a protestant, who succeeded Henry in 1547. He had become mortally ill in 1553 and attempted to remove Mary from the line of succession because of religious differences. On his death their first cousin once removed, Lady Jane Grey, was initially proclaimed queen. Mary assembled a force that successfully deposed Jane, who was ultimately beheaded. In 1554, Mary married Philip of Spain, becoming queen consort of Habsburg Spain on his accession in 1556.

Mary restored Roman Catholicism as the official religion after the short-lived Protestant reign of her half-brother. During her five-year reign, she had over 280 religious dissenters burned at the stake. Her re-establishment of Roman Catholicism was reversed after her death in 1558 by her younger half-sister and successor, Elizabeth I. After the short reigns of Elizabeth's half-siblings, her 44 years on the throne provided welcome stability for the kingdom and helped forge a sense of national identity. Elizabeth imprisoned Mary, Queen of Scots, in 1568 and eventually had her executed in 1587.

The Armada

A brief history book account of the Spanish Armada may suggest this mighty fleet was largely destroyed by a storm that developed over the English channel at an opportune time. But if we look more deeply we cannot help but be impressed by the heroism of England's defenders – from its selfless officers down to the most ragged sailor on the decks. When the tall Spanish ships kept appearing over the horizon, row upon row, the battle they faced must have seemed hopeless. But they fought anyway. We can admire the loyalty of these men to their sovereign and to their country – and find incredible their Queen's lack of loyalty to them. And we can see up close the astonishing impact an innovative weapon, properly applied, can have on a vastly superior enemy force.

Creasy describes the Armada as follows:

> *On the fleet itself the treasures of the Indian mines had for three years been freely lavished. In the six squadrons there were sixty-five large ships, the smallest of them was of seven hundred tons, seven were over a thousand, and the largest ... was thirteen hundred. They were all "built high like castles", their upper woks musket proof, their main timbers "four and five feet thick" of a strength which it was fondly supposed no English cannon could pierce ...*

> *Next to the galleons were four ... gigantic galleys carrying each of them fifty guns, four hundred and fifty soldiers and sailors, and rowed by three hundred slaves. In addition to these were four large galleys, fifty-six armed merchant vessels, the best that Spain possessed, and twenty caravels or pinnaces attached to the larger ships.*
>
> *The fighting fleet, or Armada proper, thus consisted of a hundred and twenty-nine vessels, seven of them larger than the British <u>Triumph</u>, and the smallest of the sixty-five galleons of larger tonnage than the finest ship in the English navy, except the five which had been last added to it. The aggregate of cannon was two thousand four hundred and thirty. They were brass and iron of various sizes, the finest that the Spanish foundries could produce. The weight of the metal which they were able to throw exceeded enormously the power of the English broadsides. However, the supply of cartridges was singularly meager. The king probably calculated that a single action would decide the struggle, and it amounted to but fifty rounds for each gun.*
>
> *The store of provisions was enormous. It was intended for the use of the army after it landed in England, and was sufficient to feed forty thousand men for six months. The powder and lead for small arms was also infinite.*[3]

Aboard this fleet were some 20,000 soldiers and gentlemen, six hundred priests and other people, two thousand rowing slaves, and eight thousand sailors. An army gathered from the empire embarked onboard the Armada, joined by the flower of Spanish manhood, intent on capture of England and removal from the throne of the murderous heretic Elizabeth. In Flanders, the Prince of Parma waited with an invasion army, needing only calm seas and the protection of the Spanish fleet to transport his army across the channel in flat-bottomed boats already built for the purpose.

After initial separation in stormy seas, and regrouping, the Armada sailed out from the Bay of Ferral, bound for England, on July 12, 1588.

The English Fleet

The Spanish had prepared well for the war they expected to fight. By contrast, except for a small number of well designed ships, an honest and capable yardmaster, and a few brave seamen, the English, thanks to their queen, were unprepared for any kind of war.

The Spanish leaders believed that a sizable army landed anywhere on the shores of England would capture undefended London in short order. The key to conquest would be destruction of England's fleet.

> *Quite probably they were correct. At a time like this, it would seem that the English monarch, Queen Elizabeth, would have been single-mindedly building up the English fleet. This was not the case.*
>
> *At this time the English navy did not exist as a profession. Officers, when needed, were sparingly appointed from among successful privateers, and crews were hired by the week or month or for some special service. Thirty five years earlier, in 1583, a royal commission had been appointed to look into the condition of the navy at that time. That group made strong recommendations to improve the fleet, recommendations which led to overhaul of the ships that then existed, and construction of five additional ships. The maintenance of the fleet was put in the hands of a competent sailor, Sir John Hawkins, who managed to keep the ships maintained on a tight budget that the Queen herself thought to be extravagant.*[4]

This was the same John Hawkins who had prevailed upon a penny-pinching queen in 1585 to raise seamen's wages "from six and eight pence a month to ten shillings. The increase however cost nothing to the crown, a smaller crew better paid being found to do more effective service." Creasy's account relates that Hawkins believed that with higher wages men became more healthy and self respecting "such as could make shift for themselves and keep clean, without vermin."[5]

Innovative Ship Design

The five new ships that were built as a result of the commission of 1583 used new design ideas conceived by Hawkins. There were two, eight- hundred ton vessels, the *Ark* and the *Victory*, two of nine hundred tons, the *Bear* and the *Elizabeth Jonas*, and one of a thousand tons, the *Triumph*. The last four were not completed and commissioned until early in 1588, the year the Spanish Armada sailed against

England. All five of the new English ships were of a radical new design: keels were longer, sterns and forecastles lowered, and lines sharper and finer than customary. Old-time seamen insisted they were fit only for smooth water and would sink if taken into heavy seas. A worried queen, apparently fearful that she would be proven to have paid for a foolish experiment in design, would not let Hawkins' creations go to sea, but kept them anchored in Medway harbor.

In the Spring of 1588 then, only weeks before the Spanish Armada would sail against England, the English fleet consisted of only thirty-eight ships, including the five new vessels which had not been fitted out and were not allowed to put to sea. Fifteen of the thirty-eight were small cutters and pinnaces (small sailing ships used as tenders or scouts). Only thirteen were over four hundred tons.

In 1850 the noted English historian, Sir Edward Creasy, compiled a carefully researched history of the epic battle between the English fleet and the Spanish Armada, as well as the events that preceded that event. The material which follows, provided here in some detail because of the illumination it provides, both on the subject of technological advantage and on the subject of false economies, is taken chiefly from that historical account.[6]

England's Lack of Preparedness

In 1587 Sir Frances Drake had been sent on a raiding mission along the Iberian coast with four ships, the *Bonaventure*, the *Golden Lion*, the *Rainbow*, and the *Dreadnought*. He sailed into the harbor at Cadiz, burned Phillip's shipping anchored there, and captured a Spanish galleon. When he returned to England, the crews were paid off and dismissed by order of the queen in order to conserve funds, and the ships were dismantled. Consequently, in September, 1587, at the very time Phillip was sending orders to prepare the Armada for sailing and to the Prince of Parma to prepare his invasion fleet for the Armada's arrival, there was no royal ship afloat in the channel larger than a pinnace. The English fleet was lying in harbor, rigging partially disassembled, in dockyards that were themselves feeling the effects "…like every other department of the public service from the Queens's determination to make peace." In the words of even the Queen's own favorite she "was treating for peace disarmed." [7]

Indecision

Undergoing a temporary change of heart, Elizabeth, in October, 1587, directed Hawkins to put all the available ships in condition to defend the country. The captured Spanish galleon was sold and the money used for this task. Merchant ships in port were held and crews impressed for the navy. In December she instructed Howard of Effingham to "take the ships into the channel to defend the realm against the Spaniards." Officers and apparently even crews were enthusiastic. It seemed the long wait for Phillip's hammer blow to fall was ended. A strong feeling existed among the officers that the greatest hope of success lay with a bold tactic – to dash into the Tagus and burn the Spanish fleet as it sat, before it could get underway. But before any such effort could be mounted Elizabeth immediately announced that the fleet would remain operational for only six weeks "...before the end of which the Queen confidently hoped that peace would be established."

According to Sir Edward Creasy, certain advisors had convinced her that soldiers and sailors "wished for war because it was their trade" and that Howard and Sir Frances Drake, if permitted to reach the open sea, would do something to endanger the peace negotiations upon which she relied. In addition, a rumor circulated that the Spanish fleet was dissolving. Based on that story, Elizabeth ordered half the crews that had been gathered at so much expense, dismissed, and ordered two thirds of the fleet to remain on the Thames River with only skeleton crews. The balance of the fleet was kept at harbor or within the confines of the English Channel. Phillip's peace negotiations, carried on while finishing touches were being put on the Armada, were paying off.

Whether the Spanish king shrewdly guessed what the Queen would do, or whether he received accurate intelligence reports from England, is not known. But for whatever reason, at the moment Elizabeth ordered the English crews dispersed and the fleet made for all practical purposes useless, Phillip of Spain was sending orders to the Armada to sail. Only the sudden death of one of Phillip's best admirals kept this order from being fulfilled at that critical time.

> *"Never, never since England was there such a stratagem and mask to deceive us withal as this treaty."* (Lord Howard) *"We are wasting money, wasting strength, dishonoring and discrediting ourselves by our uncertain*

Chapter 8 – Sail And Gunpowder

dallying." (John Hawkins) [8]

With the English fleet out of commission it appears possible that the Prince of Parma might even have floated his army to England without the protection of the Spanish fleet – but for the unfavorable weather that then swept the English Channel. As a result of continued unseasonable weather, the channel seas were stormy and rough, too rough for Parma's barges loaded with horses, men, and supplies.

Less than a week after ordering the fleet disbanded, the Queen became uneasy again and issued new orders, resulting in much confusion. Some ships were recalled to be paid off; shortly thereafter, others were ordered to sea, with the necessity to pay bounties and allowances to collect substitutes for the dispersed crews. To save money the Queen personally ordered the sailors' rations reduced, eliminating the traditional three times per week beef or mutton, "dictating that they defend their country and her throne on fish, dried peas, and oil." To keep the ships tied to their ports so they could not provoke the Spanish, she had even these reduced rations supplied in driblets, no more than a month's supply at a time. Supplies of other than food were also severely limited.

> *Drake had offended her by consuming ammunition at target practice. She would not give him a second opportunity. 'The proportion of powder' in the larger ships was 'sufficient but for a day and a half's service if it was begun and continued as the service might require.' In the rest of the fleet it was sufficient but for one day's service. Drake was outraged – the Queen had taken upon herself the detailed management of all supplies and expenditures.* [9]

The promising new ships, the *Triumph, Victory, Elizabeth Jonas*, and the *Bear*, were left in the shipyard, the Queen believing that "they would not be needed and that it would be a waste of money to refit them." However, when spies brought word that the Spanish fleet would sail in mid-May, she reluctantly permitted these ships to be fitted out to join the English fleet. Then as the English fleet lay waiting, prevented by storms from entering the open sea, its food supplies ran out. The sailors caught fish to prevent starvation and prayed that one of two events would occur soon: "The speedy coming of the enemy" or the arrival of the royal supply boats. The supply boats arrived first, ten days late, with provisions for only one more month, and a message from the Queen that no more would be sent.

It was not until another month had passed that she would even discuss buying more food. Then, when the food was already due to have run out, she ordered more, but the food contractors replied that it would then take four more weeks to gather the food and get it to the fleet.

Fortunately, the fleet's officers were by now accustomed to Elizabeth and had put the sailors on even shorter rations to stretch the supplies. But another problem appeared. A staple of the ration, the beer, proved to be sour and poisonous, and caused widespread dysentery. Howard and Drake ordered food and medicines under their own name for the sick men. When it was all over "the Queen called them to a sharp account for their extravagance, which has saved possibly a thousand brave men to fight for her" [10] Drake finally resolved to raid the Spanish coast to steal powder for his guns and food for his men, but bad weather, and later fear that the Spanish fleet might slip by, held him back.

The Approach of the Armada

The size and power of the Spanish fleet has been described, but one fact deserves mention: the Armada sailed without skilled pilots who might have provided a detailed knowledge of the English Channel. Although Spain carried on an extensive trade with Northern Europe, its supplies from this area were not carried in Spanish ships, but by Dutch traders. However, when the time arrived to launch the Armada, no Dutch pilots with knowledge of the channel could be found in Spain, so the fleet sailed with only a limited number of pilots at all knowledgeable of the channel and its weather.

On the nineteenth of July, 1588, the Armada reached the mouth of the English Channel. Fishing boats gave the alarm, and England know without doubt that its hour of danger had come. But to Drake, Hawkins, and Howard, the alarm came none too soon:

> *By thrift and short rations, by good management, contented care, and lavish use of private means, there was still one week's provisions in the magazines, with powder and shot for one day's sharp fighting, according to English notions of what fighting ought to be. They had to meet the enemy, as it were, with one arm bandaged by their own sovereign: but all wants, all difficulties, were forgotten in the knowledge that he was come*

and that they could grapple with him before they were all dissolved by starvation.[11]

Battle

The sea battle can be described more briefly than the train of events that led up to it. At three in the afternoon on the twentieth of July, the first ships of the Armada appeared on the horizon. As the afternoon wore on and rank after rank of the Spanish Fleet sailed slowly into view, the English sailors could see that it now consisted of some 150 ships, many of them of huge size. The two fleets remained apart throughout that night, but at dawn the huge Armada set sail toward the English ships. It was then that the Spanish commander first became aware that the conflict would not be a simple matter of closing, grappling, boarding, and capturing the English ships.

> *To Midina Sidonia's extreme astonishment, it seemed at the pleasure of the English to leave him or allow him to approach them as they chose. The high-towered, broad bowed galleons moved like Thames barges piled with hay while the sharp, low English moved at once two feet to the Spaniard's one and shot away as if by magic on the eye of the wind.*
>
> *The action opened with the Ark, carrying Howard's flag, and three other ships ... running along their entire rear line, firing successively into each galleon as they passed, then wearing round and returning over the same course. The San Matteo luffed into the wind as far as she could, inviting them to board, but they gave her their roadsides a second time and passed on ... the English were firing four shots to one, and with a fresh breeze even the galleasses could not touch them. Such artillery practice and ships so handled had never been seen ...[12]*

The rest of the English ships were following similar tactics. Throughout the morning the Spanish continued their attempts to close with the British and board, but always failed. Being downwind, their ships leaned in the water so that what shots they fired generally flew over the British ships. The Spanish could not hurt an enemy they could not touch.

As evening approached, the wind and waves rose, and the Spanish attempted to flee northward. The *Capitan* and the *Santa Catalina*, ran together in the dark. The latter was damaged so that it fell behind. Drake captured it and found several tons of powder on board, which he sent on to the rest of the fleet in the fastest British trawler available.

Back on board the *Capitan*, the officers, apparently irritated at the poor showing of their ship, began to quarrel among themselves. As the discussion became heated, the captain hit the master gunner, a German, with a stick. That individual went below in a rage, pushed a burning linstock into a powder barrel, and dived through a porthole into the sea. In the explosions that followed, hundreds were wounded, and because they could not be removed in the seas that were developing, were abandoned with the floating hulk. The English picked up the ship the next morning, took the prisoners ashore, and gave them the best (although primitive) treatment available. The English sailors, exploring the Spanish vessel, found more barrels of unexploded powder deep in its hold.

The morning of the second day was almost calm. The British, badly outsized and outnumbered, and still short of powder, resolved to "… make the best of their ability as sailors, and to wound and injure as many of the galleons as possible with least damage to themselves." A Spanish straggler was attacked. In the ensuing engagement it fired some eighty shots, high by Spanish standards, but was hit by some five hundred rounds. Nevertheless, the English had to withdraw without being able to sink it, because of lack of powder. The following day was calm. The British ships lay helpless, without powder or shot for another engagement. But, with no wind, the gigantic Spanish galleons could not move to attack. Only their smaller ships propelled by oars could attack, but they did no damage. That night, enough powder reached the English ships for one more day's fighting.

The day that would see the heaviest fighting yet dawned in time for the men on Phillip's warships to find the British towing away two of their provision ships. A group of Spanish ships gave chase, leaving a gap in the Armada. Howard could see a clear path to the enemy flagship, and resolved to sail straight for it, firing at every enemy vessel his ships passed as they penetrated the Spanish fleet. As he carried out this bold attack, a Spanish ship moved to block the way of the Ark, and, in the resulting collision, the Ark's rudder was lost. The Spanish galleons moved to close in on the disabled boat, but another British shop passed it a rope to turn it so that its sails could fill with the wind, and it escaped.

The battle raged throughout the day, the English firing and slipping away. The shops of the Armada, although doing little damage to the English, appeared on the outside to be little damaged themselves. But below decks, where the British seamen could not look, extensive harm had been inflicted. The four foot thick timber armor had acted like wooden shrapnel, severely injuring the soldiers who had been sent below for safety. Added to the physical damage was injury to the Spanish urge to fight. The Spaniards were becoming discouraged at trying to close with the rapidly moving English sailing ships, and even at their slow rate of fire, were beginning to run low on ammunition. The decisive impact of the greater speed and mobility of Howard's new hull designs was making itself felt. The Spanish Admiral sent a plea for help by fast boat to the Prince of Parma at Flanders:

> *The enemy pursue me. They fire upon me most days from morning to nightfall; but they will not close and grapple. I have given them every opportunity. I have purposely left ships exposed to tempt them to board, but they declined to do it, and there is no remedy for they are swift and we are slow.*[13]

He went on to request two loads of powder and shot, and that Parma come out and meet him. But Parma knew that his clumsy, troop-laden barges would be destroyed by the English guns and wisely remained at Flanders.

Another relatively calm day came, but the British ships had to set sail for port for supplies. Both food and ammunition were almost gone. The treasury was not empty; the royal jewels were intact, and there was powder stored in the tower, but no help was sent to the fleet. Its commanders shared among themselves the meager supplies remaining in the fleet.

The desperate English leaders sailed back out to do battle as best they could. In a rising wind and rising seas, they decided on a gamble to drive the Spanish fleet from its anchorage. Taking the oldest of their ships, they smeared the rigging with pitch, and, as dawn approached again, the sailors aboard sailed those ships directly toward the Armada, tied their rudders, set rigging and hulls afire, and dived overboard. The ships of the Armada, in their rush to escape from the path of the English fire ships, cut loose their anchors, planning to return later to retrieve them, not knowing they would be badly needed in the stormy weather that would soon hold the channel in its grip.

In a rising sea, the almost-exhausted English ships resumed the fight, sailing in as close as possible to fire, in order to conserve what little ammunition remained. The Spanish gunners were shooting from rolling platforms, and most of their shot went into the water or into the air. The English continued until evening "when almost the last cartridge was spent and every man weary with labor." [14] At nightfall the English fleet was exhausted, without food, and without ammunition. But the great Armada was decimated. Ship after ship had been sunk, and on board those still afloat, the carnage below decks was terrible. Defeated, the remaining ships fled into the North Sea. The remnants of the Armada later attempted to avoid a rematch with the British fleet and reach Spain by sailing around Ireland. Most of those who tried did not survive. The power of Spain was broken.

The Price of Victory

Nevertheless, what the great Armada could not do to the English seamen, the neglect of their Queen could. Want of food and clothing began to take its relentless toll. "A frightful mortality now went in through the entire fleet. Boatloads of poor fellows were carried on shore at Margate, and were laid down to die in the streets, 'there being no place in the town to receive them.' The officers did what they could. Howard and Drake's purses were freely opened – some sort of shelter was provided in barns and outhouses, but the assistance they could provide out of their personal resources was altogether inadequate. It would grieve any man's heart," wrote Sir Howard, "to see men who had served so valiantly to die so miserably." [15]

I am tired and sick of war. Its glory is all moonshine. It is only those who have neither fired a shot nor heard the shrieks and groans of the wounded who cry aloud for blood, for vengeance, for desolation. War is hell.

William Tecumseh Sherman

It takes 15,000 casualties to train a major general.

Ferdinand Foch

CHAPTER 9

STEAM, STEEL, AND MODERN WAR

The coming of steel-making and the steam engine in the eighteenth century led not only to the industrial revolution, but to the coming of mass warfare to the Western World.[1] Strangely enough, this development of mass warfare followed closely after a period of mutual limitation of war, a period that ended in the late 1700s. This prior time, when strict limits were placed on arms, is described by Fuller:

> *Whereas in medieval times honor existed between feudal knights, and foot soldiers took little part in battle, now honor was established between armies led by aristocrats and the common folk were excluded from the fray. During the greater part of the eighteenth century, wars were looked upon as royal games in which highly drilled soldiers were the counters or pieces, and because they were costly to maintain, armies remained small, and bloody encounters were generally avoided. The masses of the people were excluded from the struggle and depot supply replaced pillage and foraging. Further, and what was the wisest change of all, the rules of the game laid down that "neither justice nor right, nor any of the great passions that move a people should be mixed up with wars."*

Nevertheless, this age was one of great generals; for as all armies were of one model, and fire and shock were all balanced, genius, when in command, dominated the field. e.g. Charles XII, Marlborough, Eugene, Marshal Saxe, and Frederick the Great.[2]

But these days did not last. Expanding populations and the power of the machine made possible mass armies and mass destruction. "The strategy of decision replaced the older concept of limited territorial gain." [3] From there it was only a short journey to Napoleon's boast to Metternich at Schonbrunn in 1805: "I can afford to expend thirty thousand men a month." [4]

As industrialism proceeded, application of the steam engine to propel ships enabled Great Britain to extend her control over the sea, and the application of steam power to rail transport permitted Prussia and the other continental powers to support land warfare on a broad scale previously unknown in Europe.

Early Opposition to Steam Power for Ships

The paddle wheel, which had been known since Roman times, was first connected to a steam engine in the late 1700s. Robert Fulton then built the first steam-powered armored ship, The *Demologos*, later renamed the *Fulton*. [5] The first warship to be propelled by a steam engine, it was a wooden floating battery built to defend New York Harbor from the Royal Navy during the War of 1812. To protect the relatively fragile paddle wheel this ship was built with the paddle wheel placed between twin hulls, further protected by a fifty-eight inch think belt of timber armor. Its trial runs showed that something better than the paddle wheel and its bulky protective system was needed. The obvious solution was the screw propeller, invented in 1836. This invention was subsequently combined with the iron hull, which had been demonstrated on a British pleasure boat built in 1815.

The response of the British Admiralty to these promising technical innovations was much like their later reaction to breech-firing cannon, which has been described in Chapter Five. When the Colonial Office, in 1828, asked the first lord of the admiralty for a steam packet to carry mail between Malta and Greece, it received this reply:

> *Their lordships felt it was their bounden duty to discourage, to the utmost of their ability, the employment of steam vessels, as they considered that*

Chapter 9 – Steam, Steel, And Modern War

the introduction of steam was calculated to strike a fatal blow at the naval supremacy of the empire.[6]

It is not clear why their lordships felt that way. In fact, without the introduction of steam power and iron ships some thirty years later, the naval supremacy of England would have been at an end. By 1838 the combined wooden sailing fleets of France and Russia were superior to the British fleet. Only the changeover to steam and iron enabled the British to preserve their command of the sea, by use of the nation's industrial power to create English dockyards that proved to be capable of outbuilding any two of her rivals combined.[7]

The Triumph of Steam

The decision to go to iron ships fired by steam boilers did not come easily however. At the start of the Crimean War in 1853 Great Britain's main fleet was still made up of sail powered wooden ships. During the war the French built four wooden ships protected by four inches of iron plate. After the war ended in 1856 both France and England started building the world's first two armored steam warships. The British ship, *The Warrior*, was 380 feet long and weighed more than 8,830 tons. Within the space of a very few years, the end of the old navies of the world was demonstrated by the battle of the *Monitor* and the *Merrimac* at Hampton Roads, on March 9, 1862. Following that action, British Admiral Sir John Hay said: "The man who goes into action in a wooden ship is a fool, and the man who sends him there is a villain." [8]

The Steam Engine on Land

Although the War Between the States in America saw the first extensive use of rails in war, it was the Prussians who saw the war making potential of the railroad and made the first extensive test of this concept by moving a Prussian army corps of 12,000 men, with horses and guns, across country in 1846. The militaristic Prussian society appreciated the unique military value of a comprehensive national railroad network that could turn Prussia from a positional underdog sandwiched between potential enemies, into a "defense bastion in the very heart of Europe." In the early 1800s German writers were already pointing out that completion of a national railroad system would allow troops and supplies to be moved rapidly from one part of the country to another, using the advantage of interior lines of transportation.[9]

The Internal Combustion Engine on Land

Fuller states that as reach or range is the dominant characteristic of a weapon, speed and mobility in attack are the dominant characteristics of the offensive itself. For a long period of time the horse was the heart of the offensive, giving great mobility and speed of movement. But with the invention of the musket the holding power of infantry became so great that cavalry movements were immobilized,[10] As described earlier, this advantage to the defense was further increased by the use of breastworks, trenches, and repeating small arms.

Attempts had been made to armor horse as well as rider, but vulnerability to missile fire and the pike, and reduction in mobility by the weight of heavy armor made these attempts insufficient to stop the tide of change. Nevertheless, five hundred years after the disappearance of the armored knight from the battlefield, an inventor attempted to secure a patent on the concept of a war horse encased in a rigid frame, with scythe-shaped projections attached. But even though such schemes to armor-plate the horse did not prove practical, planners did not give up on the horse. Before World War I both sides confidently expected to use cavalry units armed with lances on a large scale. At the start of that conflict, Germany had ten divisions of horse cavalry and Austria-Hungary eleven. On the allied side, France had twenty such divisions and Russia forty.

Gasoline Engine plus Chain Track

In the early 1900s two inventions put the potential for mobility back into warfare: the internal combustion engine and the chain track. The power to move weight and bulk gave the power to carry armor, "and a bullet-proof horse, called a tank, was created."[11] But recognition of the fundamental importance of this new weapon was slow in coming.

> ... as its power of movement was greater than that of the foot soldier, the whole of existing military organization should have been modeled around it. Thus had the military organizer followed this pivotal idea, he would

have designed not only tanks – combat bullet-proof cross-country vehicles – but also cross country supply vehicles. He would not merely have thought of hauling artillery by means of tractors, armored or unarmored; but instead he would have mounted his guns on bullet-proof tracked vehicles. Further he would have moved his infantry in somewhat similar vehicles. In short, he would have fashioned his new model army round the internal combustion engine, armor, and the caterpillar track as armies of the muscle age of war had been fashioned around the horse, armor, and the wheel ... this was not done, because it was not realized that movement is the master element in organization.[12]

Bureaucratic Indifference

In 1914, A British army officer, attempting to interest his superiors in the concept of an armored, tracked vehicle finally did so, in the person of Sir John French, Commander of the British Expeditionary Force. He then found that Winston Churchill's Admiralty was working on just such a concept for army use, although the army did not know about it. Even more startling was the discovery that two years earlier still, an Australian inventor, E.L. de la Mole, had submitted a design for such a machine to the British War Office "where it lay unconsidered and forgotten."[13]

Despite these errors the first English vehicles of this type were eventually fabricated and prepared for shipment. To keep the concept a secret, the British shipped their primitive machines to France in large crates marked "Water Tanks", which gave them their name. The earliest "tanks" were not well designed. The engine noise made it impossible for the crew to use anything but hand signals to communicate with each other. If the vehicle tipped too far forward or backward, its gravity-fed fuel supply stopped working and the engine stopped. And when the engine stopped it was necessary to get out and crank it, a decided inconvenience on the battle field.[14]

A New Weapon Poorly Used

The first tank made its abortive battle debut at the Battle of Somme in July 1916. It use had not been well planned and any advantage of surprise was frittered away. By April of 1917 when, according to Palit, tank attack had received sufficient thought to make it possible of effective use, the Germans were no longer surprised by its appearance on the field.[15] They had developed an effective countermeasure, the land mine, and had changed their defensive tactics. They moved their main infantry forces back from the first line of trenches, let attacking forces overwhelm the first line, and push on toward the heavily manned second and third lines. The German artillery, out of reach of the allied artillery, used the area between the first and second lines, into which attacking tanks and troops had poured, as a "killing ground".

General Fuller, from the viewpoint of a tank expert, maintains that the tank was never used correctly until the fall of 1917. At that late date, with no preliminary artillery barrage to alert the defenders, groups of three tanks each were successfully sent ahead of infantry assault, resulting in "… a vast reduction of casualties for ground gained."[16] Whatever the exact date of first successful use, Palit's criticism appears justified.

> *Of the three major weapon developments in the First World War – the machine-gun, modern artillery, and armoured fighting vehicles – not one was correctly assessed by Allied military leaders for tactical employment. Not only that, they bear a heavier responsibility in that they definitely and deliberately obstructed those who more clearly saw their potential. If was because of their monumental mis-appreciation that hundreds of thousands of lives were unnecessarily sacrificed during four long years of repeated frontal assaults – all for the possession of a few hundred square yards of "no-man's land."[17]*

Tank Planning Between Wars

The interval between wars saw effective planning for tank use on the part of the Germans, and continued poor planning on the part of the English. As British tactical doctrine evolved after the First World War it culminated in the policy in effect at the start of World War II – a limited use of tanks, only to seek out and destroy enemy tanks. The Germans had a different plan, to use tanks against unarmored troops to force a quick breakthrough. Enemy tanks were to be destroyed by antitank elements and dive bombers, while German tanks were used to push rapidly though weak points and drive for strategic targets.[18]

Chapter 9 - Steam, Steel, And Modern War

The Introduction of Airborne Weapons

When World War One began, the German forces possessed some 272 aircraft, while on the other side France had 120 and England 113.

Germany also entered the war with a fleet of eighty lighter-than-air craft. But although numerous flights were made over England - some sixty in the first two years of the war - damage inflicted was apparently light, and attrition heavy. By the end of the war only seven of the ships remained.[19]

The 113 British aircraft were completely unarmed and were intended to be used only as scouts. British planners assumed correctly that they would be useful for this purpose, but assumed incorrectly that such effectiveness would go unanswered. As both sides found scout aircraft effective information sources, particularly with the introduction of aerial photography, each side independently came to the conclusion that the enemy's troublesome air scouts would have to be eliminated. Since anti-aircraft guns did not solve the problem, each side began to arm its aircraft. In the case of tractor aircraft, these were rear firing guns; in the case of pusher aircraft, forward firing. A potential victim knew then, by the type of aircraft which appeared, the particular danger he faced -- forward-firing guns or rearward-firing guns.

The Steel Wedges of Monseigneur Garros

A surprise came to the aerial battlefield when unsuspecting German pilots found themselves under attack from the rear by a single seat tractor aircraft flown by a Frenchman, Roland Garros. The Germans did not learn his secret until he was eventually forced down in German-held territory. They discovered that his gun did not fire through the propeller, it fired at it; but approximately 17 out of 18 rounds missed the spinning blade and sailed on out toward the German planes under attack. Every 18th bullet, which did hit the propeller, hit a steel wedge on its back edge, which deflected the bullet and protected the propeller.

The Fokker: Scourge of the Skies

The Germans were impressed with the ingenuity of Monseigneur Garros, but felt there

Innovation in Weapon Systems

must be a better way to accomplish the same result. In early 1915, they turned to a talented Dutch airplane designer named Anthony Fokker, who had begun to design airplanes for the German Army only after previous attempts to interest the Italians, Russians, and English in his airplanes had failed. The German request was straightforward: on a Tuesday they gave Fokker a parabellum air cooled infantry machine gun and a simple request, to figure out a way to pass a stream of bullets – 600 per minute – through the space through which the blades of the propeller also passed 2400 times each minute. By Friday, Fokker, who had never before even seen a machine gun, not only had an idea for the solution, but had built a working model. By a system of levers, cams, and pushrods, he made the propeller fire the gun during the intervals when space was available for the bullets to pass harmlessly between its blades.

This time it was the Germans who achieved a tactical surprise. Throughout the later months of 1915, the Dutch airplane with its synchronized forward firing gun made the Germans masters of the air. By the end of that year, English pilots had begun to refer to themselves as "Fokker fodder". [20] It was no doubt a cause of some consternation when the secret was revealed that the aircraft designer who built the aircraft had once tried to sell his aircraft designs to the British military and been turned down. But even more shocking was the later discovery that an English inventor had actually patented a similar device in 1914 before the war started. Unfortunately, being a loyal subject of the crown, he had immediately submitted it to the war office, where it "it was promptly filed away and forgotten." [21]

Ultimately, a German pilot flying in a fog, against official instructions, landed at a French airfield by mistake, the secret was out; and allied pilots too were soon firing through the propeller.

Following the outstanding success of Fokker's aircraft, the British government, which "could have had them for little more than the asking before the war", offered the "stupendous sum" of 2,000,000 pounds to the inventor, technically a Dutch neutral, "as an inducement to transfer his person, his knowledge, and his allegiance to the allies." However, German authorities intercepted this offer and Fokker did not learn of it until the war was over.[22]

Chapter 9 – Steam, Steel, And Modern War

Early developments in rocket propulsion

Although rocket propulsion could be discussed appropriately under either projectile weapons or aircraft, it is treated briefly here, following discussion of early use of the airplane, because its very early development seems to more closely have followed that line of thought.

Use of aircraft in war was still at a primitive state of development when World War One ended. But steady progress was made during the peace interval. But even more startling developments were on the way. P. E. Cleator relates that he himself attempted to visit the Verein fur Raumschiffahrt (Society for Space Travel) in Germany in 1934, only to find it defunct and many of its members mysteriously transferred away. He subsequently wrote to the British government to relay this important information and received this classic bureaucratic reply:

> *We follow with interest any work that is being done in other countries on jet propulsion, but scientific investigation into the possibilities has given no indication that this method can be serious competitor to the air screw engine combination. We do not consider we should be justified in spending any money on it ourselves.*[23]

Meanwhile, work continued at the German Army proving grounds at Kummersdorf and was later moved to a more isolated location at Peenemünde. The center at Peenemünde was established in 1938; and in 1939 British intelligence received an anonymous letter which revealed the fact that a rocket research center had been set up there. The letter also contained information about German radar that was later proven to be correct. Nevertheless, it was four years before the British government got around to ordering a photographic reconnaissance of the Bay of Stettin area on the black sea where Peenemünde was located. Even then, Churchill's personal scientific advisor, professor F. A. Linderman, dismissed tales of German interest in long range rockets as "a mare's nest". The Comptroller of Projectile Development at the Ministry of Supply stated authoritatively that objects on the photographs which some radicals insisted were rockets, were actually inflated barrage balloons. [24]

These men were reflecting the conviction which existed in British official circles at the time that liquid propulsion was impractical. Further events did little to improve the establishment's track record. The V-1 campaign against England was launched in early summer 1944 but with the invasion of Normandy, the British announced officially on September 7 the danger from rocket attack was then passed. On the following day, September 8, the first of several hundred V-2 rockets struck London.

Technology flows on without regard for human intentions, and each technological breakthrough offers the possibility for decisive advantages to the side that first exploits it... (But) technology itself does not automatically confer military advantage, and a blind faith in technology uncoupled with strategic analysis and deliberate participation in the technological war can lead to disaster...

Stefann Possony and J. E. Pournelle

CHAPTER 10

CHALLENGE NUMBER ONE – OVERCOMING THE POWER OF PREVAILING FASHION

A brief account of the history of weapons development and application, concluded in chapter 9 with the account of British official inability to perceive the utility of the armored, tracked vehicle, and the propeller – synchronized machine gun. This in spite of the fact that the concepts had been offered to them in 1912 and 1914 respectively. Although this may seem somewhat incomprehensible, there are two explanations why those who should have developed new weapons and ways to use them – employment doctrine in other words – failed to do so. One is predicting where advancing technology will lead. The problem of getting a clear picture of the future of technology, and the art of technology forecasting related to it, will be discussed in chapter 11. Perhaps more difficult however is the topic discussed in this chapter: overcoming the heavy hand of prevailing fashion.

While it is difficult enough to guess where technology may lead, it appears to have been even more difficult, at times, for military organizations to really want to know. There is a human propensity to overlook potential uses of new inventions, when these uses do not conform to the prevailing fashion.

Some Examples of the Power of Fashion

In an analysis prepared for a collection of predictive essays published in 1968, *Toward the year 2018*, D. G. Brennan argues that fashions are powerful determinants of the shape of warfare and weapons development, and that such fashions are usually based on "much less than detailed analysis of all alternatives." [1] He points out that a military theorist of 1918, asked to predict the course of the next 50 years, would most likely have predicted a much greater role for chemical warfare than we would think of it having today. Such widespread use of chemical agents did not occur because fashion in warfare dominated technology.

The persistent preeminence of the horse and mounted Knight in medieval European warfare was largely attributable to the power of fashion. While the cost of mount and armor restricted the dominant role in warfare to the upper classes that could afford this equipment, tradition and Papal rulings – both properly characterized as elements of prevailing fashion – often reduced the extent to which these participants did physical damage to one another. As long as combatants on both sides of any conflict obeyed this prevailing fashion, the system was relatively stable and there was strong resistance to change. Only an intrusive force from outside the system could effect such a change: and it took three successive massacres by the English Longbow – at Crecy, Poitiers, and Agincourt – to put the point across to what remained of the French nobility.

The importance of tradition

Joseph Martino illustrates that, in modern times as well, an idea or invention that promises to strengthen existing traditions or institutions is likely to be welcomed, but an innovation which appears to undermine tradition, or to alter a "society" (service or institution) away from the ideal image it has of itself, will be resisted. He describes two interesting cases involving the U.S. Navy:

> *One incident involved a proposal to develop a battleship with a single – caliber main battery. This was in fact the form the battleship did take after 1914, and in this form it was eminently successful until replaced by the aircraft carrier. Alfred Thayer Mahan, one of the foremost naval strategists*

of all time, opposed this development. He did so on the grounds that ships so equipped would fight only at long range. Since before the time of Nelson, the proper course of action for a naval vessel was to close with the enemy and fight practically at arm's length, Mahan felt that the existence of long - range weapons would create what he referred to as "an indisposition to close" and that, as a result, the Navy would lose something which had been an important tradition.

The other incident involved the steam warship Wampanoag, built shortly after the Civil War. As Morison points out, this ship was the fastest, most seaworthy steam warship build up until that time. It was literally in a class by itself. Nevertheless, after several years successful sea service by the Wampanoag a board of naval officers was convened to decide whether the ship ought to be abandoned. The board decided a number of theoretical objections to the Wampanoag existed, generally to the effect that its design did not follow common steamship practice, and therefore ought to be less cost-effective than the more conventional designs – this, in face of several years experience to prove that it was more effective! As Morison shows, however, these objections were only a cover. The real reason the board recommended scrapping the Wampanoag and the reason its recommendations were accepted was that, in the view of the board, stoking furnaces, tending boilers, and operating steam machinery were incompatible with the type of shipboard life practiced in the days of sail. In short an effective steam warship posed the risk of changing a way of life which was viewed as an ideal.[2]

Drop Tanks for Escort Fighters in World War II

Decisions that may seem somewhat incomprehensible when viewed later were not limited to the Navy. An account of the tardy adoption of drop tanks by the Army Air Corps in World War II furnishes an example. When the Army Air Corps planned a bombing campaign against Germany by the Eighth Air Force, it relied on doctrine developed by the faculty at the Air Corps Tactical School. This Air Corps policy was based on an assumption that turned out to be wildly incorrect – that bombers flying in tight formations could carry out their missions without fighter escort because their interlocking defensive firepower would defeat enemy fighters.

Consequently, plans called for sixteen heavy bomber groups and only three pursuit groups to be deployed to England in 1942. The bombers were to carry out their missions during daylight hours, using precision bombing. The British had tried daylight bombing and found losses prohibitive. But American planners were determined. To switch to night bombings would mean abandoning the doctrine on which the Eighth Air Force was based. The aircraft construction program in the U.S. was based on this preponderance of heavy bombers.

A raid by 38 bombers over the French Coast resulted in no loss of bombers but the bombers were protected by 400 fighters, mostly British. Inexplicably, a memo circulated in Washington afterward seemed oblivious to this latter fact and proclaimed that heavy bombers in strong formations could be employed without fighter escort. In reality, the combined bomber offensive envisioned by Army Air Corps planners was unattainable; Eighth Air Force bombers "scarcely dared to penetrate the borders of Germany". Flight distances were at least 250 miles, beyond the range of the P-47 aircraft available, which had a combat radius of approximately 175 miles. When the Eighth Air Force attempted missions beyond the range of escort fighters

> *... the loss rate climbed sickeningly. During the first six months of 1943 when few deep penetration missions were attempted, bomber losses on each mission nevertheless ran at six or seven percent. The damage rate ranged from 35% to 60%, which meant that follow-up strikes often had to be delayed while repairs were carried out.... By July of 1943 it was becoming evident the British predictions were all too true; unescorted bombers were taking prohibitive losses. A raid on Hamburg suffered a twelve percent loss. A raid against Kassel lost twenty-three percent; another strike, against Kiel, lost fourteen percent ...Losses were beginning to run ahead of replacements. The Schweinfurt raids in August and October 1943 suffered twenty-six-percent and thirty-three-percent losses. The latter figure came close to the forty percent casualties of the infamous Charge of the Light Brigade.[3]*

Drop tanks for fighter aircraft would later enable fighters to accompany bomber aircraft on raids deep into enemy-held territory. But none were available in 1943. Why was this? At least a partial answer to this question can be found in a paper by I. B. Holley, Jr. in the record of a historical conference on Air Leadership held at Bolling Air

Chapter 10 – Challenge Number One – Overcoming The Power Of Prevailing Fashion

Force Base April 13-14, 1984.

> *Looking back with the perspective of the historian, it is obvious that extending the range of fighter aircraft so they could escort bombers for the most distant targets was vital to the success of strategic bombardment. ... If drop tanks made a crucial difference one may well ask why were they not employed much earlier. Why were there no long-range fighter escorts in 1942? Had they been available the combined bomber offensive would have been a much more adequate test for the concept of strategic bombardment much sooner. Such a test might have reduced the scale and duration of the invasion by ground forces – even if bombing could not entirely obviate that effort (as some enthusiasts in the bomber community had suggested might be possible). Historical investigation of the problem has uncovered a number of curious facts.[4]*

The Idea of Using Drop Tanks Was Not New

In the early 1920s, Carl Spatz – later the first Chief of Staff of the U.S. Air Force – commanded the first pursuit group. At that time he declared that bombers should be escorted by pursuit aircraft with the range equal to that of bombers. Billy Mitchell urged a similar course of action. However, in March 1941, when a proposal to use drop tanks on fighter aircraft was submitted to headquarters, it was disapproved by the Plans Division in the Office of the Chief of the Army Air Corps. The memo of disapproval, signed by General Spatz who by now was chief of the Plans Division, stated that such tanks would "provide opportunities for improper tactical use of pursuit types." Holley discusses this memo in some detail:

> *The very language of this memo is revealing. Note the phrase "it is believed that," construction much employed by staff officers. It expresses an opinion not a fact. ... While Spaatz signed the document and therefore accepted responsibility for its contents, the memo was drafted by one of his staff whose initials were on the file copy Hoyt S. Vandenberg, the officer who some years later would follow Spatz as Chief of Staff of the newly formed post – war Air Force.[5]*

Had both men adopted a position that heavy bombers could penetrate all defenses without escort? Gen. Holley notes that in April 1943 General Spaatz wrote, " I am just as convinced as ever that the operations of day bombers if applied in sufficient force... cannot be stopped by any means the enemy now has and ... recent raids have gone a long way to demonstrate that fact to the more persistent unbelievers." Holley points out that the opinion expressed in the first part of that sentence became fact in the second part which might suggest that the general's faith in the strategic bomber may have taken precedence over his objectivity.[6]

It Can Be Difficult to Think About New Things.

New technology is important but to maximize the effectiveness of new systems it is vital to develop new doctrine – a plan for the best ways to use the new technology. Employment doctrines are often held hostage to an organization's culture. One practical way to summarize an organization's culture consists of three sentences: (1) This is who we are. (2) This is what we do. (3) This is what we stand for. Historical events have proven that a military organization's culture will exert a strong influence on the way it sees itself going about the business of fighting.

> *An organizations's culture specifies what is of primary importance to the organization, the standards against which its successes and failures should be measured. ... In short, the culture is the essence of what is important to the organization.*[7]

Whether it was the continued dominant role for the armored, mounted knight confidently expected by nobility on the Continent, confidence in the continued primacy of the longbow by the British while gunpowder weapons were being developed on the Continent, or the assumed usefulness of horse cavalry by both sides planning the makeup of their forces prior to World War I – it is evident that each of these groups thought the future would be much like the past.

> *It is difficult to imagine a future that does not closely resemble the present. Most of us see events as forming a trajectory through time, starting with the past and extending through the present and into the future. Within this view, the present follows so logically from the past that it hardly seems that it could have been any different. This same logical determinism seems to ensure that the future is a projection of the present.*

Chapter 10 – Challenge Number One – Overcoming The Power Of Prevailing Fashion

> *Somehow, it is difficult to think that this smooth extrapolation could have a zig or zag in it, that it is not foreordained to be what it is going to be.*[8]

Organizations shaped to promote internal stability and organizational efficiency may find it difficult to respond to new threats and opportunities – a situation called the Strategic Paradox.

> *Top managers usually rise to preeminence by successfully executing the established strategy of the firm. Thus they may have an emotional commitment to the status quo and are often unable to see things from a different perspective. In this sense they are a conservative force that fosters inertia.*[9]

Two factors will bias the ideas of mangers (leaders, directors, planners, 'top brass') concerning what they should monitor in the environment and how to interpret what they see. The first is what they are conditioned by their experience – and their organization's culture – to think is important. The second is the success of the firm's current strategies. The more successful the organization is, or has been recently, the less likely they are to spend time and effort scanning the exterior environment for the possible threats and opportunities that may result from advances in technology.

Bureaucratic organizations that emphasize top down authority structures tend to be particularly good at resisting innovation. Peace-time military organizations combine a traditional rigid top down authority with a bureaucratic organizational structure. What has been lacking? A willingness to question institutionalized systems, tactics, and employment strategies.

Fashion versus technology in the Modern Era

Brennan suggests a more recent example of the power of fashion.

> *Some fashions are ambiguous or negative in their effects. For example: it became standard doctrine in the United States in the 1950s to rely for security against Soviet attack on the threat of destroying a large fraction of the Soviet economy and population – that is, on deterrence by something like "massive retaliation." There was no apparent technical alternative in the 1950s or early 1960s, so it was not then merely a fashion. But after*

> *1966 it began to appear that active defense against attack by ballistic missiles could achieve a degree of effectiveness not previously seen... Some students of strategy have argued that the fashion (as it is now appropriate to call it) of relying on a large number of Soviet "hostages" has become so ingrained in the American establishment that... It is not acceptable to shift strategic emphasis toward defense and away from deterrence.*[10]

Implicit faith in this particular fashion forms the underlying assumption behind much of the material currently written on the subject of defense spending and force planning. But Possony and Pournelle, in their book *The Strategy of Technology: Winning the Decisive War*, insist that this fashion is no different from similar persistently-held beliefs in the history of warfare. It can expose its followers to the possibility of unpleasant surprise. They put it this way

> *"... Technology has a habit of being richer than even the most imaginative planners predict. Technological breakthroughs in missile defense are inevitable, and we must be a position to take advantage of them... Defense technology is at the moment in the lower left hand quadrant of the technology curve. We see no reason why it should not continue to a breakthrough.*[11]

Indeed, since those words were written, developments in sensors and microelectronic computing power, miniaturization, and cost reduction have been greater than might have been guessed at. "Brilliant Pebbles" furnishes a prime example.

> *Brilliant Pebbles was a part of the Strategic Defense Initiative (SDI) in 1989. The idea was first pitched in the early 1980's and it was called "smart rocks." It was re-named "Brilliant Pebbles" in 1988 by the Reagan Administration and it has been known as that since. It was a non-nuclear, space-based, boost phase anti-missile program. The concept was to deploy 4,000 satellite constellations in low-earth orbit that would fire watermelon sized projectiles made of tungsten to intercept targets.*[12]

It was a concept for a space-based weapon that would target objects – in this case missiles in the boost phrase – from an altitude that conventional weapons could not reach. It would however be vulnerable to anti-satellite weapons. In the late 1980s

Chapter 10 – Challenge Number One – Overcoming The Power Of Prevailing Fashion

development of the technology proceeded rapidly and the first Bush administration supported it. George Monahan, director of the Space Defense Initiative announced that it would be the U.S. missile defense system to be deployed first. His successor pared the contractor team to two companies TRW-Hughes and Martin Marietta. All that was needed was funding.

In 1991, following several years of inner turmoil, the Soviet Union imploded. Despite the end of the Cold War, Brilliant Pebbles remained an essential part of the U.S. missile defense architecture. That same year, computer simulations demonstrated that, if it had been deployed during the Persian Gulf War, Brilliant Pebbles would have shot down every Scud missile launched by Saddam Hussein, including the salvo attack on Riyadh, Saudi Arabia. Following the Middle East crisis, Brilliant Pebbles was enhanced to give its interceptors the ability to enter into the atmosphere, thus improving its overall effectiveness against Scuds and cruise missiles.

In 1993 the Clinton Administration eliminated the Brilliant Pebbles program through a series of budget cuts. The technology itself would continue to be tested for a short time: one year later, NASA launched a deep-space probe known as "Clementine," which had been built using first-generation Brilliant Pebbles technology. Clementine successfully mapped the entire surface of the Moon. The mission, which cost $80 million, effectively "space qualified" Brilliant Pebbles' hardware. However, the Clinton Administration did not take any steps to resurrect the program.[13]

The Need for Continued Innovation

History has demonstrated that possession of a weapon system that works extremely well will give its possessor an advantage. But that advantage will be temporary. Progress in technology soon makes that cherished weapon system obsolete. The more astonishing its success the greater the tendency of its developers to depend too long on both the system and the strategy they developed to use it. A similar thing can happen to great corporations that polish the process of making and marketing their mainstay product at the same time progress in technology is working to make that product suddenly obsolete. Eastman Kodak and the advent of digital photography furnish one example There, jobs, whole communities, and the savings of investors are at risk. When military organizations do this, lives are at stake. The consequences can be sudden and disastrous. The French mounted Knight encountering the English Longbow was an example, and, in turn, the English archer who suffered terribly when he was sent out to face French cannon at Formingy in 1450. Previous sections have described the fate of the impressive but clumsy floating fortresses of the Spanish Armada in 1588, and the plight of bayonet-wielding infantrymen thrust into the barbed wire and machine gun environment of World War I. A dominant weapon can create an aura of invincibility .

However, dominant weapons do not last forever. The 'latest thing' eventually becomes old.

I say technically I don't think anybody in the world knows how to do such a thing (build a nuclear – armed Intercontinental ballistic missile). And I feel confident it will not be done for a long period of time to come.

> Vannever Bush, chairman of a
> special committee on new weapons
> for the Joint Chiefs of Staff, 1945.

CHAPTER 11

CHALLENGE NUMBER TWO: LOOKING ACCURATELY INTO THE FUTURE

Dominance Is Important, But Always Temporary

The first Prussian breech-loading infantry firearm demonstrated that even an incremental advance in weaponry can confer a critical advantage in battle. That incremental improvement led to a change in tactics, reloading while remaining in a prone position, that had a powerful psychological effect on opposing Austrian troops. Rapid Prussian conquest of Austria strengthened the Prussians against the French in the Franco- Prussian war of 1870. Greek fire or sea fire helped the Eastern Roman Empire maintain control of its vital sea lanes from the seventh through the ninth centuries. For over 100 years the English Longbow assured the supremacy of British field forces over those of France. Then a French innovation, cannon, drove the English from their fortifications in Franc. Britain used cannon from foundries built by Henry VIII mounted on innovative faster ships designed by Sir John Hawkins to defeat a seemingly superior Spanish Armada. That victory allowed England to survive as a nation and permitted the rapid development of the Plymouth and Massachusetts colonies in the 1600s – colonies which would form the nucleus of a nation that would emerge in the New World in the 1700s.

We have seen however that nations which cling too long to a technology that once gave them great advantage are often leapfrogged by rivals who move on to a more superior technology

The present importance of technological advantage

The Byzantine Empire lasted 1123 years – roughly a thousand years after Rome fell. It possessed both commercial strength and to a consistent, logical approach to war and alternatives to war. It developed better weapons than those possessed by its opponents – and strategies and tactics to match.

The Byzantines faced a succession of would – be conquerors who wished to eliminate the Eastern Empire as a commercial and military force in the ancient world. Is the U.S. in a somewhat analogous situation? Possessed of commercial, industrial, and technological advantages, but relatively manpower short, it is been threatened since World War II by an assortment of governments that don't like us very much. These governments have been quite willing to expend the lives of their people in limited war after limited war. Unlike the careful Byzantines, however, from 1950 through the early years of the 21st century America has become involved in a succession of conflicts with two common characteristics: They have ended without U.S. victory and have been extremely costly in American lives. These costly "little wars" were fought on the fringes of the basic East – West conflict that Possony and Pournelle see as chiefly technological. They put it this way:

> *The new form of warfare has its roots in the past, but it is a product of the current environment. World War II was the last war of industrial power and mobilization, but it was also the first war of applied science. The new war is one of the directed use of science... Wars of the past were wars of the attrition of the military power which was a shield to the civilian population and the will to resist. The new technology has created weapons to the be applied directly and suddenly to the national will.*[1]

We know U.S. supremacy does not bring on global war or wars of conquest, because this country held an "absolute mastery" during the period of its nuclear monopoly. There is no assurance, however, that Russian or Chinese superiority would see that

Chapter 11 – Challenge Number Two: looking Accurately Into The future

power so restrained. The possible results of US superiority and inferiority are contrasted:

The winner the technological war can, if he chooses, preserve peace and order, act as a stabilizer of international affairs, and prevent shooting wars. The loser has no choice but to accept the conditions of the Victor, or to engage in a shooting war which he has already lost.[2]

Five Common fallacies about technology

Let us examine five common fallacies concerning technology. As Will Rogers put it, "It isn't what we don't know that gives us trouble, it's what we know that ain't so."

One common fallacy is that the march of technology can be halted by agreement. Although one side of a potential conflict might possibly be restrained by good intentions or an inability to spend adequate sums to secure better weapons, it seems unlikely that both would. More to be expected would be that both sides, unless very foolish, would pursue better technology. (More about the march of technology later.)

Closely related to the first fallacy is the second: Assumed symmetry of motives or actions, or as it has been put: "The enemy won't do what we won't do." ("Why should he do that?") "The enemy is not doing what he is, in fact, doing." ("He can't be that stupid, and it isn't cost effective.")[3]

A third fallacy is the idea of "overkill". Since the onset of the era of Mutual Assured Destruction, this mantra has been advanced regularly by domestic proponents of unilateral arms reductions. In brief it asks, "If the U.S. has enough missiles to destroy (pick a percentage) of an enemy's high value targets why have more?" But this line of reasoning ignores three facts: (1) as time goes on, age can lower an equipment's reliability, (2) the technological race can make previously invulnerable offensive forces quite vulnerable, and (3) enemy action (a surprise counterforce strike for example) could destroy many U.S. systems. The idea of "strategic sufficiency" as it has been called should be considered in light of these possibilities.

The best protection against losing one's second – strike force to enemy preemption is constant updating of the force.[4]

A fourth fallacy would maintain that if it has been constructed, it's obsolete. This fallacy would lead one to believe that prototype and research are all that are needed. This should have been thoroughly discredited by the experience of the French in 1939. The French Army had, for some time, prototypes of aircraft and armor of better quality than those possessed by the German Army; but since even better weapons were coming, the French did not put these existing weapon types into the inventory. "While they waited for the best weapons, they lost their country." [5]

The fifth fallacy is fear of obsolescence – which is based on acceptance of the fourth above. "Why build it if it will soon be obsolete due to the march of technology?" This assumes that technological advances in the military arts are automatic. While it is doubtless true that technology does have a momentum which cannot be stopped, the direction and timing of developments can be drastically changed. An opponent who elects to swim with the stream, rather than drift with it, can make the seemingly small improvements in the early breakthroughs that lead to first place in the inevitable contest for military advantage ("there is no prize for second place in combat.").[6]

The difficulty of making accurate predictions

Even unhindered by prevailing fashion, we humans find it difficult to predict future technology developments. A high level government study undertaken in 1937 missed all the important developments which took place in the following 20 years: jet engines, radar, inertial guidance of missiles, use of data processing computers, the use of earth satellites for communication and navigation, and nuclear weapons. [7] Predictions made since World War II have continued to illustrate the inability of even the most respected of authorities to predict, with any certainty, that particular weapons can or cannot be developed, or, if developed, will be of significant impact. During the hydrogen bomb controversy, one school of scientists was convinced that the weapon could never be built and insisted that even if it could, it could never be transported.

> *Another group believed the opposite and was proven to be correct. As it happened, they stood at the threshold of the sharply rising portion of the technological S – curve.*[8]

We are reminded of the statement which introduced this chapter, a statement made in

Chapter 11 - Challenge Number Two: looking Accurately Into The future

testimony before the special Senate committee on atomic energy in December, 1945, by a respected authority who had been chairman of a special committee on new weapons for the Joint Chiefs of Staff. Vannevar Bush dismissed as impossible the concept of an Intercontinental ballistic missile.[9]

In his text on technology forecasting, Joseph Martino presents four interesting, but inaccurate, predictions of the future made by authorities in the field. These are listed in Table 11-1.

Table 11 - 1

Who said it and the prediction

Franklin Roosevelt then assistant Sec. of the Navy, 1922:

"The day of the battleship has not passed, and it is highly unlikely that an airplane or fleet of them could ever sink a fleet of Navy vessels under battle conditions."

Simon Newcomb, prominent astronomer, shortly before the first flight of the Wright brothers:

"The demonstration that no possible combination of known substances, known forms of machinery, and known forms of force can be united in a practical machine by which man should fly long distances through the air, seems to the writer as complete as it is possible for physical fact to be."

Professor a. W. Bickerton:

"The foolish idea of shooting at the moon is an example of the absurd lengths which vicious specialization will carry scientists working in thought - tight compartments... The proposition appears to be basically impossible."

Thomas Edison:

"My personal desire would be to prohibit entirely the use of alternating currents. They are dangerous... I have always consistently opposed high - tension and alternating systems of electrical distribution... Not only on account of danger, but because of their general unreliability and unsuitability for any general system of distribution."

Technology Forecasting

What are the possibilities of some new technology? Is it possible to select areas of technology that have potential? Perhaps the best we can do is make intelligent guesses. Fortunately, application of the methodology of technological forecasting can help. To better understand technology forecasting it is necessary to discuss four areas:

First – We need to be aware of the importance of supporting technologies.

Second – We need to understand the phenomenon known as the Growth Curve.

Third – We must be award of the tendency of technical people to shift from optimism to pessimism as they look farther into the future.

Fourth – we will want to think about how to answer the unanswerable question "Is invention inevitable?"

First – Supporting technologies.

It is accepted as a commonplace truth that one advancement in technology leads to another, but it may be as accurate to say that one technology often waits on another. It is held up until a key discovery is made in some supporting technology. As summarized by Joseph Martino:

> *A technology cannot exist in isolation, but only as part of an interlocking web of technologies that provide it with support.*[10]

The steam engine was introduced in the British Isles in 1705, but remained without significant size or power for nearly 60 years. Machinists could not bore large cylinders with enough precision to allow use of tight-fitting Pistons. Wilkerson's first precision boring lathe appeared in 1774 and made powerful steam engines possible. With this new lathe machinists could drill cylinders up to 5 inches in diameter. The steam engine then became a useful prime mover.

(Delayed development of the steam engine has been attributed by other writers to a conspiracy of sorts, a desire by British authorities to hold back industrial and technological developments.)[11]

More recently, the advent of powerful computers, accompanied by development of complex mathematical models of the laws governing the combination of particular atoms to form molecules has led to "dry lab" experiments to discover new chemical compounds and processes. In the area of missile weapons, commonplace fiber glassing technologies developed for civilian use have been used in rocketry. A significant increase in ICBM accuracy occurred between 1964 and 1968, not through deliberate application of new technology, but through reduction of international geophysical year data, which increased understanding of gravitational anomalies, the largest single factor in the ICBM error budget.[12]

Something that seems small can become a key element combined with related inventions. The bridle, bit, and stirrup, rather small things in themselves, turned something that had been around for a long time into a master weapon. Invention of the percussion cap and integrated fulminated detonators made possible repeating, breech loading firearms and a revolution in field tactics. Still later, solid state electronics made possible an effective Minuteman missile.

There is an interesting sidelight in the story of the Minuteman. It needed a lightweight and accurate guidance system. It was lightweight but at first, unfortunately, not accurate.

> *...when Minuteman II was deployed the reliability of its guidance and control system was about 1/6 of requirement. It took three years to overcome the difficulty, but then performance exceeded specifications, if we had been attacked during this period, we would have been in a fine mess. Since the mishap was widely rumored, the Soviets probably knew about it – fortunately the USSR lacked adequate strength.*[13]

Second – Identification of promising areas of technology: The technological growth curve, or S – curve.

Technological forecasting has become a discipline in itself, a discipline which can be applied to assess the degree of advancement that a particular area of technology may achieve, and to screen areas of possible technological development, to help choose those where significant advancement appears likely. A key to understanding technical progress is the growth curve. Although some technologies have exhibited linear or other forms of technical progress, many analysts believe that growth in capability of various technological devices or systems often exhibits the same behavior as the growth of biological organisms. It can therefore be best represented by one of a family of "S" curves. Figure 11-1 shows the general nature of a growth curve. Figure 11 - 2 shows the basic nature of such S - curves in comparing two illumination sources. Figure 11-3 shows curves for two technologies that were competitive for a time, with the one starting later, initially inferior, eventually surpassing the first. [14] Note: These growth curves are based on Dr. Martino's examples in his text *Technology Forecasting for Decision Making*.

Three things are apparent in figure 11 - 2: (1) the capability of both incandescent and fluorescent - Mercury vapor lamp technologies did grow in an S-curve over time. (2) the total illumination area, incorporating both technologies also grew in the same fashion. (3) the earliest and most primitive fluorescent devices had to nearly equal the most highly developed incandescent light sources to be competitive

At a given point in time two competing systems may have roughly equal capabilities but it is possible that the earlier system is based on a mature technology high on its S-curve. The second system, with limited capability at that time, may be progressing very slowly because it is still on the first segment of its growth curve. If based on technologies still in their infancy, it may promise explosive growth in capability. Decisions based on optimistic "blue - sky" predictions for new and unproven technologies should be avoided, but we need to be aware of the potential for rapid growth in new technologies - and the growth that may come because of progress in supporting technologies.

Chapter 11 – Challenge Number Two: looking Accurately Into The future

Size of Pumpkin,
Performance Capability,
Etc.

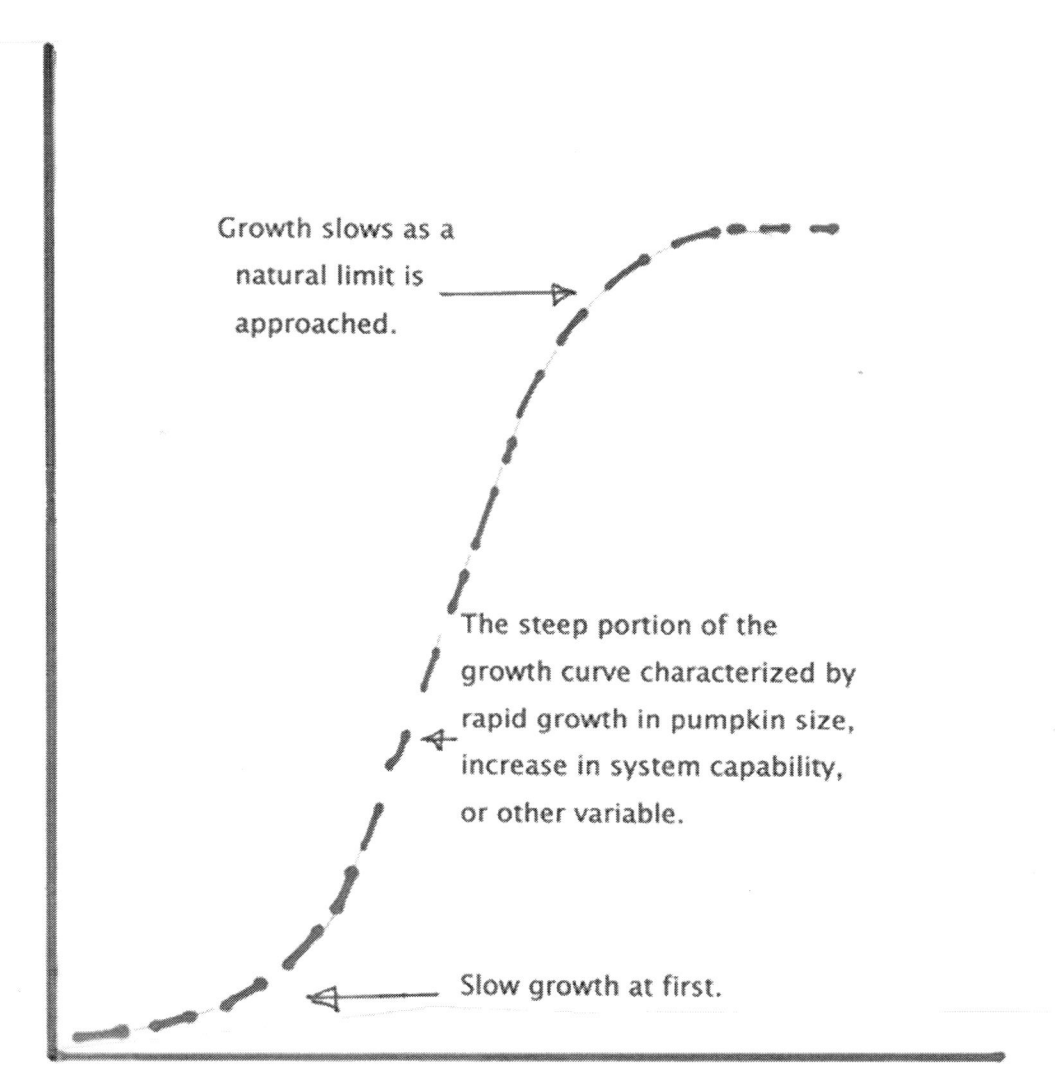

Growth slows as a natural limit is approached.

The steep portion of the growth curve characterized by rapid growth in pumpkin size, increase in system capability, or other variable.

Slow growth at first.

Time, Research Effort Over Time, Etc.

Figure 11-1

Innovation in Weapon Systems

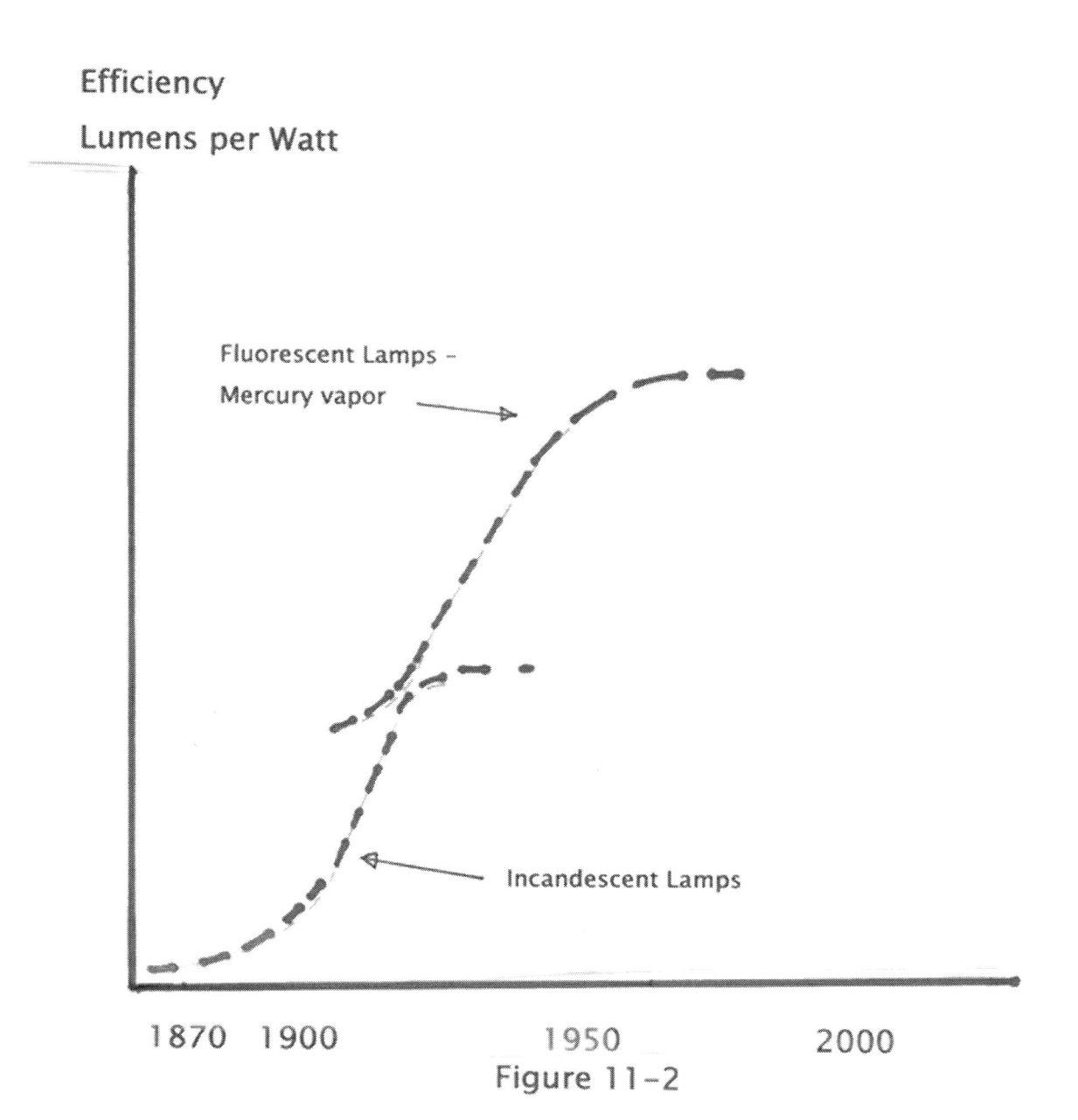

Efficiency
Lumens per Watt

Fluorescent Lamps – Mercury vapor

Incandescent Lamps

1870 1900   1950   2000
Figure 11-2

Chapter 11 – Challenge Number Two: looking Accurately Into The future

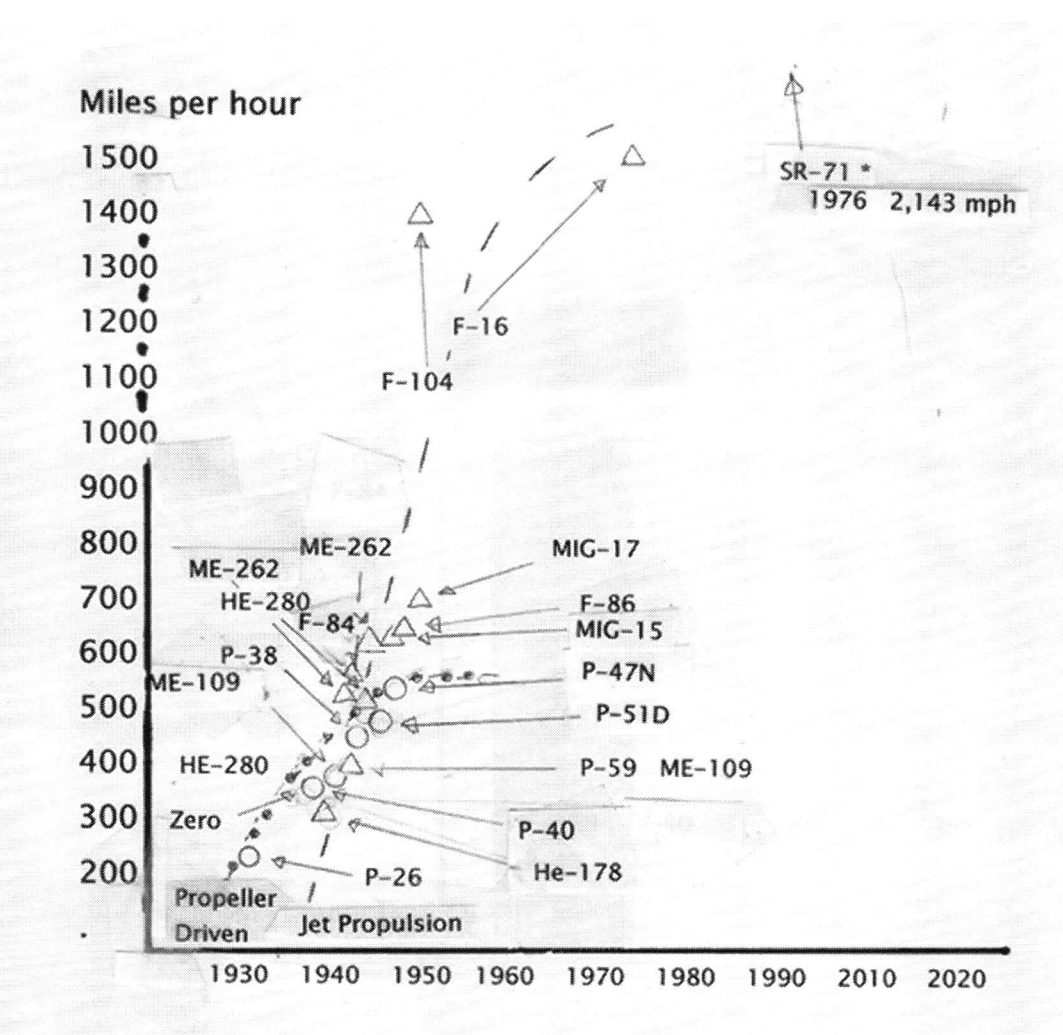

Figure 11-3

Note: Idea for graphs through 1950 generally based on Dr. Martino's graph, page 108, *Technology Forecasting for Decision Making*. Data points for aircraft identified on this graph are from Wikipedia. *As a special purpose aircraft perhaps the SR-71 should be considered an outlier. Yet it did demonstrate the capability of technology.

Third – The shift from optimism to pessimism

Studies have proven the existence of a systematic shift from optimism to pessimism as the time length of a forecast increases. Many forecasters tend to be optimistic in the short run, especially about work they are familiar with or are responsible for. They are optimistic about overcoming near term obstacles and typically under-estimate the resources required to do so. In the long run, however, they tend to become very pessimistic, seeing many barriers that at the moment they do not know how to overcome. Martino's rule of thumb is that it the shift from optimism to pessimism comes at a point about five years in the future.[15]

Martino maintains that forecasting is not limited to a particular technology.

> *It appears also that reasonably accurate forecasts can be made not just of technical parameters, such as speed or other technical description, but of a total capability, such as air transport productivity (which is a product of both speed and size of the transport) and that this can be done without the forecaster having to "invent" the specific technology to bring about the improved capability. It has been objected that the forecaster cannot project his forecast beyond the limit of the current technical approach without knowing what the successor technical approach will be... However... It is possible to forecast beyond the limits of current technological approaches, without knowing what the successor technical approach will be."* [16]

In this regard Dr. Martino points out that the growth of jet aircraft speed (figure 11-3) was actually a continuation of the trend already established by propeller driven aircraft.

Fourth – Is invention inevitable?

Martino indicates that a relatively strong case could be made for the thesis that invention is inevitable, although he would himself modify that proposition somewhat. He discusses the work of W. F. Ogburn published in *Social Change* in 1922. Ogburn collected and listed 135 duplicate and independent inventions and scientific discoveries that took place between the early 1600s and late 1800s – from fields which included astronomy, mathematics, chemistry, and physics – and argued that the discoveries were inevitable. By this he did not mean that society could do away with scientists, or that a society could not suppress inventions. His point was simply this:

> *There is a certain rate of occurrence of hi – level scientific talent, and that if this talent is occupied in scientific and technical activity, individual*

Chapter 11 – Challenge Number Two: looking Accurately Into The future

inventions and discoveries come inevitably.[17]

Many examples of multiple, independent inventions do exist. Fulton's Clermont was preceded, in the 20 years before its successful operation, by a rash of other "more or less successful steamboats" operated by their inventors in both the U.S. and England in the 1900s.

> *In our own century Norbert Weimer put the finishing touches on a mathematical theory of prediction and time series that was later employed in the design of radar controlled antiaircraft guns in World War II. However a Russian mathematician (Kolmagorov) had independently developed the identical theory at approximately the same time.*[18]

In a similar vein Martino describes the discovery of nuclear fission as "a history of near misses and repeated experiments" and calls attention to the striking parallelism between US and German atomic research in its early stages.

> *Up to the point where the German government for all practical purposes abandoned its project, the activities of the two projects were essentially identical, with similar solutions being devised at about the same point in time. Even after the end of practical attempts to devise a weapon, the German theoretical efforts closely paralleled those in the United States, with similar mathematical techniques being devolved independently and simultaneously. Weinberg and Nordheim, two American scientists who examined captured German documents after the war, even raised the ethical question of whether it was proper for American scientist to publish their own work without acknowledging that German scientists had achieved similar results independently and simultaneously.*[19]

Are inventions then inevitable? Is the march of technology to be considered to have a life of its own? Will certain inventions surely be invented by someone somewhere? These are important questions. Dr. Martino summarizes several contrasting views on the inevitability of invention:

> *E. E. Morrison – surrounding circumstances can practically force invention or discovery: Years, centuries, of experiment with windmills and waterfalls had produced by the 18th century a very sophisticated technology to go*

with these sources of energy. In some countries, impressive linkages to transmit power three or four miles from the waterwheels had been developed. All the elaborate machinery needed was a more effective prime mover. Since there wasn't a steam engine it became necessary to invent it.

S. C. Gil fill proposed a "principle of equivalent invention: Perceived needs are met by various unlike as well as duplicate solutions, so that any great invention is simultaneously paralleled by other, often utterly dissimilar, means for reaching the same end of the same time, e.g. Reaching California by clipper, steamer, pony express, railroad, and telegraph... Hence, no single invention ever revolutionizes civilization, nor brings, simply through having been invented, any important changes in the life of men.

J. Hacke – In a study of the economic and social impact of the transistor pointed out that many other inventions came at approximately the same time as the transistor, which performed the same functions, but were not quite competitive with the transistor. The growth of electronics might have continued, much as it did, without the transistor, but it would have been based on a number of completely different devices.

J. Schmookler – who argues against a deterministic view of technology: Schmookler discusses the advent of the automobile and points out that it was not the inevitable result of technological growth. It was just as much a product of an individualistic society that valued the prestige, flexibility, privacy, and utility it could provide. Consequently, the automobile has not revolutionized life in a much less individualistic Soviet Union. As for its technological base, its development did not depend on the internal combustion engine since steam and electric power plants were highly competitive in the early days of the automobile.[20]

Martino says in fact that a practical steam car could have been built in the early 1800s if there had been any market for it and concludes that it came when that market existed. Even a century earlier cruder technology could have been made to serve.

Chapter 11 – Challenge Number Two: looking Accurately Into The future

Invention A can eliminate the need for Invention B.

One invention can eliminate any need for the development of a different invention to do the job performed by the first. And even the timing of some invention, and who discovers it, can affect subsequent history. For example, nuclear fission could have been discovered two or three years before the experiments by Hahn and Strassman in Berlin in 1938. Had this occurred during the rearmament of Germany, when the German government was still willing to look at new technology, Germany might have developed an atomic bomb before 1939. [21] The impact on the world of an atomic blitzkrieg is difficult to imagine.

For invention to lead to new weapons a sufficient technology base must exist. When it does, and a need exists, similar inventions may come with a rush. Exactly what specific advancements and inventions will come on the scene will depend on the highly fluid reality of human events. A specific invention or even its functional equivalent will not inevitably come about.

> *A change in the timing of one major invention may so change things that some inventions may never be made, while others become desirable and are therefore made... It is indeed possible to make a fairly plausible case for the view that invention is inevitable, and that breakthroughs come practically on schedule, however this view attempts to claim too much. While it may be true that necessity is the mother of invention, and inventions are inevitable when the necessity arises, there is nothing inevitable about necessity. Even fairly minor alterations in the course of history could create quite different sets of "necessities" which will therefore change the breakthroughs supposed to come "inevitably."* [22]

Different systems and different technologies compete for attention and funds. Selection of the best is necessary, but difficult.

In war, the morale is to the physical as three is to one.

Napoleon Bonaparte

In technological war, organization and leadership is to morale as six is to three.

Porsonny and Pournelle

Other things being equal, battles are won by superior technology, but clearly superior technology prevents battles.

Porsonny and Pournelle

CHAPTER 12

TODAY'S REALITY AND TODAY'S FASHION

This chapter presents data from published sources concerning U.S. strategic forces and doctrine for their employment. It recalls the tendency illustrated in previous chapters for a country's success with one system and a doctrine developed for it to cause that country to 'go to sleep' so to speak, while some other country develops the next big thing. It examines the relative strength and modernity of the U.S. strategic force and those of potential aggressors and asks if those trends point to an insecure future for our nation.

Trends in U.S. Defensive Forces
Versus Potential Rivals

An attempted comparison in military strength between the United States and possible rivals on the world stage can get hung up on at least four snags: (1) System capabilities for the U.S. are unknown and could not be discussed here if they were. The proper course is to rely on published information. (2) Comparative information on foreign systems is likewise unknown. Once again, it is necessary to rely on published data. (3) Whatever information is current at the time of this writing could quickly become obsolete. (4) The world of national defense involves land, sea, space, and aeronautical systems and is composed of both tactical and strategic systems. In order to keep this comparison manageable it is restricted almost entirely to what are commonly called strategic systems, those capable of striking an opponent's homeland.

The U.S. Depends on Deterrence.

According to press sources, the U.S. has only a minimal defense against ballistic missiles.

The Soviet government came to the conclusion early on that defense of Moscow was a good idea and started development of a missile defense system in 1971. Currently called the A-35 anti-ballistic missile system, it has Gorgon and Gazelle missiles with nuclear warheads to intercept incoming ICBMs.[1]

The U.S. has installed an interception facility in Alaska. Once known as National Missile Defense it now has the less imposing title of Ground-Based Midcourse Defense and has reached initial operational capability with only a few interceptor missiles. Unlike the Russian system it does not use nuclear warheads but instead launches a kinetic projectile to destroy an incoming warhead by impact. After the ninth of September 2011, then-President Bush cited the proliferation of ballistic missiles as reason for building it. It has the limited goal of shielding the continental U.S. from a rogue state. Its location would make it less effective against an attack from the Middle East but the site would be favorable In the event of an attack from North Korea.[2]

The U.S. Cold War Buildup and Its Demise

As brutal and perhaps unreasonable as it sounds, the only thing preventing an attack on the U.S. by another nuclear power is the threat of nuclear retaliation - a sort of "If

Innovation in Weapon Systems

you attack us, we'll kill you" threat. (A more detailed discussion of deterrence will be found farther on in this chapter.)

The U.S. developed a strategic "triad" of land-based missiles, sea-based missiles, and aircraft to assure the Soviet Union that the U.S. could carry out such a threat. It deployed three strategic aircraft: the Supersonic B-58, the B-47, and the giant B-52. It built liquid fueled Titan and Atlas intercontinental ballistic missiles (ICBMs). Today, all these systems are gone except the latest versions of the venerable B-52. It also fielded the Minuteman solid fueled ICBM in three versions, Minuteman I, Minuteman II, and Minuteman III, each more modern than the last. Minuteman III was followed by the LGM-118 Peacekeeper. It was hoped that both Minuteman and Peacekeeper missiles would be protected from a counterforce strike (a first strike by an enemy attempting to wipe out the U.S. nuclear force) by being placed underground in hardened silos.

The LGM-118 Peacekeeper was first known as the MX (X for experimental). It could carry 10 re-entry vehicles, each armed with a 300-kiloton W87 warhead in a Mk.21 reentry vehicle (RV). It was deployed at the Strategic Air Command, 90th Strategic Missile Wing at the Francis E. Warren Air Force Base in Cheyenne, Wyoming in re-fitted Minuteman silos. Under the START II treaty the missiles were to be removed from the US nuclear arsenal in 2005, leaving the LGM-30 Minuteman as the only type of land-based ICBM in the arsenal. The START II treaty never went officially into effect; nevertheless the last of the LGM-118A Peacekeeper ICBMs was decommissioned on September 19, 2005. It was planned to move some of the W87 warheads from the decommissioned Peacekeepers to the Minuteman III. [3]

In an example of swords into plow shares, a private launch firm, Orbital Sciences Corporation, developed the Minotaur IV, a four-stage civilian expendable launch system using old Peacekeeper components.[4]

Today, in accordance with agreements with the Russian government to limit the number of delivery vehicles and the warheads on each, the U.S. for the land-based missile portion of its triad relies solely on 550 Minuteman III missiles.

As an important part of the U.S. nuclear deterrent the U.S. Navy developed missiles carried by submarines. The first was the UGM-27 Polaris. The UGM-73 Poseidon missile was the second. It was replaced by the Trident I in 1979, which in turn was replaced by the Trident II in 1990.[5]

The UGM-133A Trident II, or Trident D5, is carried on 14 Ohio class boats and four British Vanguard-class submarines – 24 missiles (now 20) on each *Ohio* class and 16 missiles on each *Vanguard* class. The Trident D5LE (life-extension) version will remain in service until 2042.[6]

New Russian Missile Submarine

The Russian Federation has developed a new Borei class submarine. Three have been completed and are active, four more are under construction, eight are planned. It is known variously as the Dolgorukiy class; sometimes transliterated as Borey, and as the Dolgorukiy class after the name of the lead vessel, *Yury Dolgorukiy*). The class is intended to replace the Delta III, Delta IV and Typhoon. It is named after Boreas, the North wind. The Severodvinsk-class submarine is the new Russian attack submarine

A replacement for many types of submarines, the Dolgorukiy-class submarines are slightly smaller than the Typhoon-class in terms of length, 560 ft as opposed to 574 ft. The crew, at 107 people is also smaller.[7]

Is Our Concept of Deterrence Shared by the Russians?

Brent Scrowcroft pointed out in 1985 differences between American and Russian ideas of deterrence. He noted that the concept of deterrence is not new – George Washington having said that the best way to prevent war is to prepare for war. He makes the point that, for the west, deterrence was nothing more than what the then-Soviet leaders believed it to be and that we should try to understand what they believe it to be, not what we believe it to be. As Scrowcroft said:

> ...from what can be gleaned from Soviet literature, force posture, force deployments, and concepts of force employment, there is little evidence that the Soviet Union subscribes to the concept of assured destruction.[8]

According to Scrowcroft, Soviet leaders had a more traditional view of strategic warfare: nuclear weapons might be decisive but not revolutionary. They tended to think more in terms of war-fighting rather than just deterrence

A nation which, for example, practices the lengthy process of reloading missile silos in military exercises can hardly be expected to be sympathetic to the notion of assured destruction or, as it is sometimes described, deterrence through pain ... Instead, the Soviet Union is likely to be deferred under extreme circumstances only by the belief that she cannot achieve her objectives/ that is, deterrence through denial.[9]

A high level study has pointed out that the official military doctrine of the Russian General Staff still views the United States as a potential enemy. Russian nuclear programs and strategy appear to be a continuation of Cold War attitudes, with a dangerous assumption that the use of a small number of nuclear weapons will end a conflict with the Russians victorious. The first use of nuclear weapons is part of Russian military doctrine, something that goes beyond the declared policy of any other nuclear power. This low nuclear weapon-use threshold is linked to the old Soviet view of the world and Russia's role in it. Adding impetus to this policy: the fact that Russian dreams of grandeur about its historical superpower role cannot be supported by the Russian economy.[10]

Some writers insist that its nuclear capability is the only claim Russia has to world power status. Would it then be willing to respond to perceived threats by lowering the nuclear threshold? Colonel-General Vladimir Yakovlev, the head of Russia's Strategic Rocket Forces said exactly this in an interview in 1999: "Russia ... is forced to lower the threshold for using nuclear weapons, extend the nuclear deterrent to smaller-scale conflicts, and openly warn potential opponents about this."[11]

According to the Russian press, Vladimir Putin urged the continued development of nuclear weapons while he was a cabinet member in the Yeltzin government. In his present position as head of the Russian government he has continued to show this interest.

President Putin and Defense Minister Ivanov have personally participated in strategic nuclear exercises. Such participation will certainly continue in the future. In August 2005 President Putin not only participated in a strategic nuclear exercise but actually flew in a Russian Blackjack bomber which launched four of the new Russian KH-555 long-range land attack cruise missiles. Western Presidents would simply never act in this manner.[12]

Chapter 12 – Today's Reality And Today's Fashion

Nuclear Treaties between the U.S. and the Soviet Union/Russia

SALT I There have been a number of treaties between the U.S. and the Soviet Union (and later the Russian Federation). The first was called SALT I. The acronym stood for Strategic Arms Limitations Talks. Over a six month period ending in May 1972 agreement was reached to limit strategic interceptors number to 200 (the ABM or Antiballistic Missile Treaty) and to put restrictions on offensive missiles. Neither side was to build any new ICBM silos or increase their size "significantly". The agreement limited the number of sea launched ballistic missiles and the number of subs carrying them. The ICBM limit was set at 1,607 silos and the limit for SLBM launching tubes was 740. The ABM limits were later reduced to 100. The agreement did not limit numbers of warheads and did not address the number of strategic aircraft. Possibly as a result of progress in missile defense technology and a desire to protect the American population from attack by a smaller or rogue nation, the U.S. withdrew from the ABM Treaty.

SALT II Negotiations began in November 1972 for a follow-on treaty and SALT II was signed in June 1979. Both the U.S. and the Soviets were limited to 2,250 delivery vehicles. A delivery vehicle was either an ICBM silo, a SLBM launch tube, or a heavy bomber. The Soviets at that time were over the limit and would have had to reduce by more than 200 delivery vehicles. The U.S. was under the limit and could have built up. When the Soviet Union invaded Afghanistan in 1979 the Carter administration did not submit the treaty to the senate. Although it was never signed both sides subsequently pledged themselves to honor it. When he took office, President Reagan felt that the treaty was contrary to U.S. interests and declared it would not govern future decisions.

START I President Reagan proposed a new treaty in the early 1980s and the Strategic Arms Reduction Treaty (Start I) was finally signed in July 1991. It called for reductions in delivery vehicles (to 1,600) carrying a limit of 6,000 warheads. It required information exchange, on-site inspection, and use of satellite photos. Reductions were completed in 2001. The treaty expired in 2009.

START II As a result of agreement between George H. W. Bush and Boris Yeltsin, more talks were begun and another treaty, START II was signed in January 1993. It would not have required the destruction of warheads but would have required scrapping of some delivery vehicles. Ratification was delayed; a protocol agreed on in 1997 moved the effective date from 2003 to 2007. However the Soviet Duma put conditions on its approval, and the treaty never went into effect.

START III Bill Clinton and Boris Yeltzin settled on a framework for START III, which would have required warhead destruction but negotiations were never begun.

SORT In 2002, George Bush and Vladimir Putin signed a Strategic Offensive Reductions Treaty (SORT), also called the Moscow Treaty. It did not require destruction of warheads or delivery vehicles. The objective was to limit warheads to 1,700 – 2,200 for each country.

New START The current treaty, signed in April 2010, is called New START. It limits each side to 1,550 strategic nuclear warheads. They are to be deployed on no more than 700 delivery vehicles, which apparently can be ICBM silos, SLBM submarines, or heavy bombers. It also limits deployed and non-deployed launchers to no more than 800. The treaty does not put limits on defensive systems or long range conventional (non-nuclear) systems. It includes on-site inspections, satellite photography, and telemetry exchange on missile tests. It went into effect in February, 2011 and is to be effective for ten years.[12]

Treaty limits under "New START" are summarized below.[13]

Type	Limit
Deployed missiles and bombers	700
Deployed warheads (RVs and bombers)	1,550
Deployed and Non-deployed Launchers (missile tubes and bombers)	800

The United States began implementing New START reductions even before the treaty was ratified. This required removal from service of at least 30 missile silos, 34 bombers and 56 submarine launch tubes, though missiles removed would not be destroyed and bombers could be converted to conventional use. Four of 24 launchers on each of the 14 ballistic missile nuclear submarines would be removed but none of the Ohio-class subs would be retired.[14]

The treaty places no limits on tactical systems.

Mobile Rail Systems

New START places no limits on railway mounted systems (missile trains) but such launchers would be covered under the generic launcher limits. Inspection details would have to be worked out between the parties. Russia has professed concern about U.S. desires to develop a precision, non-nuclear, quick strike capability (an ongoing effort to develop a capability to deliver a precise conventional weapon strike anywhere on Earth within one hour) and has used this as justification for development of a Russian rail mounted system (The Soviets once had one but it used the SS-24 missile which various press reports indicate has been taken out of service or is in the process of being replaced.) The Russian Defense Ministry reported to the Kremlin that it will move ahead with an updated missile launching system using rail (a "missile train") Because of its size, Russia has great strategic depth, almost guaranteeing the survivability of its ICBMs after a hypothetical first strike.

> *This was the logic that drove the Soviet-era development of the SS-24. The SS-24 program consisted of 14 silo-based ICBMs and 42 rail-based platforms with a range of 6,200 miles and a capability to deliver up to 10 warheads to separate targets via a multiple independently targetable reentry vehicle capability (MIRV).*[15]

Christopher Ford of the Hudson Institute has pointed out that, in his view, the language of the New Start treaty opens the way for development of a Soviet missile train if they wish to do so because -for mobile missiles - a missile "launcher" is defined by the treaty as the erector-launcher mechanism "and the self-propelled device on which it is mounted". A transporter-erector on a rail car that is pulled by a railroad engine is not "self-propelled". Ford points out that silo locations are known and wartime reloading might not be feasible. However, reloading of mobile systems - either road mobile or rail mobile - even in wartime could be possible and "wouldn't take very long." According to this line of reasoning the treaty would appear to allow either side to have as many rail-mobile launchers as they liked, as long as the launchers were not loaded even though the missiles themselves were quickly accessible. Even a loaded erector might not count, since it is not self propelled.

The enormous Soviet SS-24 missile, a MIRVed giant with 10 warheads similar to the U.S. Peacekeeper, was designed to be either silo-launched or rail mobile. The SS-24 was subject to elimination under the Start agreement, but that treaty, of course, is no more and if you can build road-mobile systems like the SS-27, rail-mobile applications sound quite simple indeed.[16]

U.S. Strategic Forces Have Been Reduced in Numbers

To get an idea of trends in strategic armaments and reductions in numbers of U.S. forces we can look back at numbers from approximately thirty years ago. The Reagan administration published figures for both the U.S. and the Soviet Union in official annual reports. Press reports today give numbers for the U.S. (in accord with the New Start limits outlined above) but information on Russian forces today lacks the detail presented in those government reports.

Strategic Forces in 1985

Each table lists missiles in apparent order of modernity.

(Note: This data is taken from Soviet Military Power – 1985, Superintendent of Documents, U.S. Government Printing Office, pages 27-38 with supplementation from Wikipedia. Quotations are from Soviet Military Power unless otherwise identified.)

U.S. Intercontinental Ballistic Missiles in 1985

Name	No. deployed	Warheads/missile	Launch
Peacekeeper	(in development)	(up to 10)	Hot
Minuteman III	550	1	Hot
Minuteman II	450	1	Hot
Titan II	26	1	Hot

Chapter 12 – Today's Reality And Today's Fashion

Soviet Intercontinental Ballistic Missiles in 1985

Name	No. deployed	Warheads/missile	Launch
SS-X-26	(nearing deployment))	1	Cold
SS-X-24	(in development)	10	Cold
SS-19	360	6	Hot
SS-18	308	10	Cold
SS-17	160	1	Cold
SS-16	(unknown)	1	Cold
SS-13	60	1	Hot
SS-11	520	1-3	Hot

Notes: The SS-18 and SS-19 are very large missiles with heavy throw weight. The SS-18 was designed to destroy ICBM silos in the United States. The ten multiple independently targeted warheads were each approximately 20 times as powerful as the Hiroshima and Nagasaki devices. The SS-18 force had the capacity to " destroy more than 80 percent of U.S. ICBM silos using two nuclear warheads against each."

U.S. SLBMs in 1985

Name	No. deployed	No. of warheads Per missile	Launch
Trident C-4	(unknown)	8	Cold
Poseidon C-3	(unknown)	10	Cold

Soviet SLBMs in 1985

Name	No. deployed	No. of warheads Per missile	Launch
SS-NX-23	(in flight test)	(unknown)	Cold
SS-N-20	(unknown)	6–9	Cold
SS-N-18	(unknown)	1–7	Cold
SS-N-17	(unknown)	1–2	Cold
SS-N-8	(unknown)	1–2	Cold
SS-N-6	(unknown)	1–2	Cold

Even in 1985 the Soviets had the world's largest ballistic missile submarine force. "As of early 1985 the force numbered 62 SSBNs carrying 928 nuclear tipped missiles." Missile range was such that two thirds of this force could remain in waters near the Soviet Union, or even in their home ports, and still hit the U.S. with their missiles. (Note: The Typhoon class submarine (being replaced by the Borei class)) was approximately one and a half times the length of a football field – including the end zones and longer and wider than the Ohio class subs, still the most modern possessed by the U.S.)

In the areas of both land-based and Sea Launched Ballistic Missiles, it is apparent that the Soviets displayed a propensity to build a greater variety of and larger missiles.[17]

Current U.S. Strategic Forces

Summing up U.S. Strategic forces is simple. New START Warhead Limit – 1,550 distributed among 500 Minuteman III missiles[18], fourteen Ohio-Class subs carrying up to 20 missiles[19] (reduced from 24 to 20 by New START) with others distributed among an estimated 100 heavy bombers ready for action on any given day from among 76 aging B-52 Stratofortresses, 63 B-1 Lancers, and 20 B-2 Spirits.[20]

Chapter 12 – Today's Reality And Today's Fashion

Current Russian Strategic Forces [21]

Land-based

Older missiles still in service: SS-18, SS-19, and the SS-25. Gradually being phased out. Phase out predicted to be completed by 2022.

Replacement systems: SS-27 Mod 1, and the SS-27 Mod 2. Each of these is built in both silo-based and road-mobile versions. Deployment of the Mod 1, which carries only one warhead, was completed in 2012. The Mod 2 carries multiple independently-targeted reentry vehicles (MIRV). Deployment began in 2010.[22]

Sea-based

At the start of 2015 the active SLBM force contained three types of submarines: Delta III, Delta IV, and one Borei class (others not yet on line).

Delta III – 2-4 (published estimates vary) apparently based with the Pacific Fleet. Each carries 16 MIRVed missiles.

Delta IV – 5-7 (published estimates vary) apparently based with the Northern Fleet. Each carries 16 MIRVed missiles. High operational costs and retirement of the *Typhoons'*s SS-N-20 missiles caused some *Delta III's* to be reactivated.

Typhoon – Russia built several Typhoon-class SSBNs apparently based with the Northern Fleet. None are now operational although one was used as a testing platform for a new missile. The Russian Navy canceled a Typhoon modernization program in 2012, after concluding that modernizing one Typhoon would be as expensive as building two new Borei-class submarines.

Borei-class – 3 (Eight total in current plans – but press sources indicate more may come.) [23]

Air-based

According to open source estimates, Russia has two heavy bombers, 13 Tu-160

Blackjack, 29 Tu-95MS Bear H6, and 30 Tu 95 Bear H16 for a total of 72 aircraft. Another estimate places the total at 66. There are at least two varieties of air launched cruse missiles available and a large number of gravity bombs. The Blackjack can carry twelve of either type of cruise missile, the two versions of the Bear can carry six (H6) and 16 (H16) of one type.[24]

Hot Launch versus Cold Launch

In what is known as a "hot" launch, a rocket blasts from its below-ground silo using its rocket engine. We could expect this to destroy the silo. A missile that is cold launched is blown from its silo – in the manner of a sea launched ballistic missile launched from a tube on a ballistic missile submarine – after which the rocket engine ignites when the missile is in the air. Presumably, the U.S. relies on satellite photography of missile fields (counting the number of silos) as a final check of Russian adherence to limits on launchers. If some Russian ICBMs are cold launched, leaving silos reusable – perhaps after refurbishment – this would seem to open up a way for possible use of numbers of missiles in used silos considerably above treaty limits. As early as 1985 the Soviet Union appears to have been preparing for reloading of mobile launchers.[25]

> *The belief that a nuclear war might be protracted has led to the USSR;s emphasis on survivability along with war reserves... and the capacity to reload launchers...Plans for the survival of necessary equipment and personnel have been developed and practiced. Resuppply systems are available to reload SSBNs in protected waters.*[26]

In contrast, it can be seen that all U.S. land-based missiles were of the hot launch variety. Obviously the missile's silo would be destroyed upon launch and thereafter unusable. If satellite imagery were used for verification purposes it would seem to be a simple task for a foreign power to verify the number of silos and therefore the number of missiles possessed by the U.S. Obviously this would not be the case with the Soviets/Russians. If each silo were reloaded only once, they could have twice as many silo-launched missiles as U.S. analysts, relying on space imagery to count the number of visible silos, would assume.

Chapter 12 – Today's Reality And Today's Fashion

Trends In Modernization

Warhead Modernization

The Federation of American Scientists reports that Russia is modernizing its strategic and nonstrategic nuclear warheads. According to the Federation's figures the Russians possess 4,500 nuclear warheads. Approximately 1,780 are deployed on missiles and at bomber bases. Seven hundred strategic warheads and approximately 2,000 nonstrategic warheads are stored and not deployed.

The FAS reports that that an estimated 311 ICBMs can carry around 1,050 warheads. A process of retiring Soviet-era ICBMs is underway. ICBMs removed from service are replaced by the SS-27 Mod 1 (Topol M), the SS-27 Mod. 2, two follow-on versions of the SS-27 still being developed, and a new liquid-fuel "heavy" ICBM.[27]

> *Russia's upgrades ... raise questions about Russia's commitment to its obligations under the nuclear Non-Proliferation Treaty to reduce and eliminate nuclear weapons.... The U.S. State Department released aggregate New START numbers from the 1 March 2016 data exchange. Russia declared 1735 deployed warheads, 521 deployed launchers, and 856 total launchers. In September 2015 the numbers were 1648, 526, and 877 respectively.... Some older missiles were apparently withdrawn from service....The U.S. numbers in March 2016 were 1481 warheads, 741 deployed and 878 total launchers (1538, 762, and 898 in September 2015).[28]*

China

According to a report in the *Bulletin of the Atomic Scientists* China currently has 45 missiles that can strike the continental U.S. The bulletin points out that China is assigning a growing portion of its warheads to strategic missiles and apparently has equipped some with multiple independently targetable re-entry vehicles (MIRVs). Older liquid fuel rockets are being replaced with solid fuel, quick launching missiles. One liquid fuel rocket, the DF-SA, a two stage, silo based system, has apparently been targeted at the U.S. and Russia since the early 1980s. The Bulletin quotes March 2015 testimony by Admiral Cecil Haney, head of Strategic Command, to the Senate Armed Services Committee saying that China was "... conducting flight tests of a new mobile missile, and developing a follow-on mobile system capable of carrying multiple warheads."[29]

Russian Modernization and New START

An April 2016 article by long-term defense analyst Bill Gertz summarized strides made by the Russian Government in its Intercontinental Ballistic Missile Program. Under the headline "Russia Doubling Nuclear Warheads: New multiple-warhead missiles to break arms treaty limit" this article provided a number of specifics. Gertz maintains that the Russian government is doubling the number of warheads on new missiles, using multiple independently targetable reentry vehicles (MIRVs) on two newly deployed systems, the land based, road-mobile SS-27 and the sub-launched SS-N-32, and this has put Russia over the limits set by the New Start treaty.

> *The 2010 treaty requires the United States and Russia to reduce deployed warheads to 1,550 warheads by February 2018. "The Russians are doubling their warhead output," said one official. "They will be exceeding the New START (arms treaty) levels because of MIRVing these New systems."* [30]

The State Department revealed in January that Russia currently has exceeded the New START warhead limit by 98 warheads, deploying a total number of 1,648 warheads. According to the article, while the U.S. is currently below the treaty level of 1,550 warheads the State Department indicated in January that the Russians are now over the limit by 98 warheads. There was conflict between agency statements – the State Department saying that it has the right to inspect these new missiles, the Department of Defense saying that, although the 2010 treaty gave the U.S. the right to check warheads, Russian officials are attempting to block U.S. weapons inspectors from doing so.[31]

More Modern Launch Vehicles

In 2009 President Obama declared his objective of a world without nuclear arms and world leaders held a meeting in Washington in April 2016 to discuss nuclear security. Notably, Russian's Vladimir Putin did not attend.

> *Russia, however, is embarked on a major strategic nuclear forces build-up under Putin. Moscow is building new road-mobile, rail-mobile, and silo-based intercontinental-range missiles, along with new submarines equipped with modernized missiles. A new long-range bomber is also being built.*[32]

In contrast, after seven years of defense reductions, the Pentagon appears hard-pressed to find funds to modernize the aging U.S. nuclear forces. Adm. Cecil Haney, commander of the U. S. Strategic Command, told Congress in March there was concern about the Russian modernization concern.

> *"When you look at what they've been modernizing, it didn't just start," Haney said. "They've been doing this quite frankly for some time with a ... crescendo of activity over the last decade and a half."* [33]

A spokesman for the State Department's arms control, verification, and compliance bureau says that the numbers are simply the result of normal fluctuations due to modernization before the deadline for treaty limits, February, 2018, when each country is to have no more than 700 deployed treaty-limited delivery vehicles and 1,550 deployed warheads. On the other hand the Pentagon insists Russia has upgraded its nuclear doctrine and Russian officials have threatened a number of times that their military might use nuclear arms against the United States.

A former Pentagon official involved in strategic nuclear forces, Mark Schneider, has warned that Russia is not reducing its nuclear forces. Older SS-25 road-mobile missiles have been taken out of service but new SS-27s with multiple warheads have taken their place. This official went on to say that the administration has been deceptive about the benefits of New START.

> *The administration public affairs talking points on New START reductions border on outright lies. The only reductions that have been made since New START entry into force have been by the United States. Instead, Russia has moved from below the New START limits to above the New START limits in deployed warheads and deployed delivery vehicles.*[34]

Former Secretary of Defense William Perry has supported the idea that the United States should unilaterally eliminate all its land-based missiles and rely instead on nuclear missile submarines and bombers for deterrence. However the logic of this seems obscure because he goes on to say that "I highly doubt the Russians would follow suit" by eliminating their land-based missiles.[35]

If Russia withdraws from the New START arms accord it will have had six and a half years to prepare to violate the treaty limits, at the same time the United States was reducing it forces to the treaty limits. As Schneider puts it:

> *Can they comply with New START? Yes. They can download their missile warheads and do a small number of delivery systems reductions," "Will they? I doubt it. If they don't start to do something very soon they are likely to pull the plug on the treaty. I don't see them uploading the way they have, only to download in the next two years.*[36]

Other Countries

The Russian Federation, discussed at some length above is not, of course, the only nation state possessing nuclear weapons. The Federation of American Scientists published in 2016 a list of countries and the number of warheads they were thought to possess.[37]

Country	Warheads
U.S.	6,970
Russia	7,300
United Kingdom	215
France	300
China	260
India	100 – 200
Pakistan	110 – 130
Israel	80
North Korea	<10
Total	**15,350**

Great Britain currently maintains four submarines with 16 Trident missiles each but is thinking about disarming. At any given time France keeps one sub on patrol. It also has the capability for aircraft delivery. China apparently has multiple delivery options. It is slowly increasing warhead output. After having breached the non-proliferation agreement, India is expanding delivery options and increasing its warhead stockpile. Officially, Israel has no nuclear capability. As for North Korea, it is unknown what delivery options they may have.[38]

Missile Defense Systems

The Federation of American Scientists has found that Chinese leaders are giving serious thought to deployment of a system to defend against ballistic missile attack.

> *Unlike some years ago, there is little doubt today that China is developing a strategic BMD capability; their flight tests alone make that clear," said authors Bruce MacDonald and Charles Ferguson, who spoke with more than 50 Chinese and American experts, including Chinese officials, military officers and academics. While Chinese BMD is in the development stage, it does give Beijing the option of deploying a missile defense capability – or not – depending on its assessment of the international situation.*
>
> *"At a minimum, it appears that a Chinese deployment of strategic BMD is probably less unlikely than most U.S. defense analysts have in the past assessed," the study said.*
>
> *"Given the extended duration of China's strategic BMD development program, going back two to three decades, it is safe to say that China is not on any crash course to develop, much less to deploy, a strategic BMD system. Nonetheless, China's program has reached a stage of maturity that gives it a viable option to deploy if it so chooses.[39]*

The Federation of American Scientists report, "Understanding the Dragon Shield: Likelihood and Implications of Chinese Strategic Ballistic Missile Defense," was prompted by a lack of public analysis of what impact a Chinese ballistic missile defense might have on the United States. Since a thin defense would not be effective against the U.S. arsenal, the main target of Chinese missile defense would be China's neighbors.[40]

Innovation in Weapon Systems

Unmanned Aerial Vehicles

The Global Hawk, as one example of Unmanned Aerial Vehicles, utilizes satellite data links for control and transmission of images to its operators far away. In addition to its basic utility, this aircraft is important for two reasons. First, it is a true innovation.

Although this can be said to some degree with every incremental capability leap in combat aircraft, the Global Hawk is not just another fighter or surveillance aircraft, it is an entirely new concept of operations, and what we learn from it, both the bad and the good, will find its way into even more advanced unmanned technologies of the future.[41]

Second, the Remotely Piloted Vehicle concept was considered impractical for years because of a data link problem. In a sense, the near earth satellite, along with miniturization of computer power, represents an enableing technology that permitted development of this innovation. Near earth satellites have value not only as "eyes in the sky" but in support of systems such as this, as well as communications, and GPS satellite positioning devices. In time of war survival of one's own satellites and damage or elimination of an opponent's satellites could assume great importance– not unlike the conclusion reached by both sides in World War I that the opponent's troublesome air scouts should be eliminated.

Satellites, data links, and anti satellite weapons

The Russian government has been successful in developing an anti-satellite system. Bill Gertz reported in December 2015 that a third test of an anti-satellite system in November, following two failures, was a success. This direct-ascent system has been given the name "Nudol" a term that has been mentioned in various reports over the last few years [42]

The tests thus far have not involved a kill vehicle – only tests of what is probably a solid fuel rocket. Russia now joins China as one of the only two nations with strategic space warfare weapons. In October, China conducted a flight test of its anti-satellite missile, the Dong Neng-3 direct ascent missile. In the past, State Department officials have characterized China's development of space weapons as destabilizing. "The continued development and testing of destructive [anti-satellite] ASAT systems is both destabilizing and threatens the long-term security and sustainability of the outer space environment," Frank Rose, assistant secretary of state for arms control, verification and compliance, said in February.[43]

China has conducted a number of tests of anti-satellite weapons. A test in 2007 left thousands of pieces of debris that continue to threaten manned and unmanned satellites.

Mark Schneider, now with the National Institute for Public Policy indicates the Russian test points to a failure of the United States to prepare for possible war in space. According to a February 2015 unclassified Defense Intelligence Agency report to the Congress "Chinese and Russian military leaders understand the unique information advantages afforded by space systems and are developing capabilities to deny US use of space in the event of a conflict.

> *There is an enormous asymmetry in play regarding space weapons, said Schneider,. For decades the Congress has prevented the US from putting weapons in space and even developing a ground-based ASAT capability," Schneider said. "There is no such constraint upon the Russians and Russia violates arms control treaties when this is in their interest to do so and they find ample opportunity to do this.*[44]

For the second time in less than a year a polar-orbiting weather satellite has generated orbital debris but this appears to have been the result of natural causes. The Joint Space Operations Center (JSpOC) said it detected the breakup of a U.S. National Oceanic and Atmospheric Administration satellite that had been retired in 2014 and was tracking an unspecified number of "associated objects" in the orbit and added that it did not believe the debris resulted from a collision with another object, but that NOAA 16 broke up on its own.[45]

Summary

The public press tells us that the number of U.S. strategic systems has been reduced and is stable. There are reports that Russia is modernizing both launch vehicles and warheads. There have been suggestions that the Russian Federation could be preparing to break through New START treaty limits in both, but this deals with intentions, and can be only informed speculation at this point and not known fact. As noted earlier a high level Russian official has warned that the Federation might use nuclear weapons in lower level conflicts. The head of the Russian government has demonstrated a continuing interest in the potential of nuclear weapons even to the extent of riding in a Russian bomber in a military tactical exercise.

The U.S. depends on satellites for a number of uses – some of which could be vital in a time of hostilities. Russia, and China both are testing direct ascent anti-satellite weapons. Published reports indicate the U.S. has no such systems. Russia has an anti-ballistic missile defense system around Moscow. China is conducting tests of such a system. The U.S. had installed a limited system that would presumably be useful against a small number of missiles from North Korea, but, because of its location, not so useful against an attack from the Middle East. It has abandoned a promising space based anti-ballistic missile system, "brilliant pebbles", cutting off funds for the program.

The Russian military is improving the technology of its offensive missile systems with reported potential to break out of New START treaty limits and improving its anti-ICBM and anti-satellite technologies. The Chinese military, although numbers are small at present, is modernizing and increasing the number of its offensive missile force and testing anti-ICBM and anti-satellite technology.

America continues to rely for defense of the homeland on the seventy year old doctrine of Mutual Assured Destruction.

There are few die well that die in a battle.

William Shakespeare, Henry V

CHAPTER 13

THE IMPACT OF THE NUCLEAR BOMB

This discussion would be incomplete if it did not include the impact of the first use of atomic weapons.

The Battle That Never Was

Use of the atomic bomb abruptly ended World War II. On July 26, 1945, the Potsdam Ultimatum had called on Japan for unconditional surrender. Three days later, on July 29, Japanese Premier Suzuki issued a statement in which he scorned the ultimatum as "unworthy" of official notice. Eight days later, on August 6 the first atomic bomb was dropped on Hiroshima. Three days later, on August the second was dropped on Nagasaki. The next day, August 10, Japan declared its intention to surrender. On August 14 Japan accepted the Potsdam terms.

The Purple Heart

The Purple Heart is awarded to American troops wounded in battle and to the families of those killed in action. In 2000, the government ordered a new supply of Purple Hearts. The old supply, manufactured thirty five years earlier in advance of the planned invasion of the home islands of Japan, had begun to run low.[1]

Scholars who have examined captured Japanese documents and those who interviewed Japanese military men after the war have given us a sense of what that invasion would have cost. Had the invasion taken place, there would have been a medal shortage.

U.S. Leaders Planned to Invade Japan to End the War

U.S. military officials planned for invasion of the Japanese Home Islands in the fall of 1945. They had estimated that many soldiers and marines would be killed in a beachhead invasion but examination of documents and interviews with Japanese military people after the war proved those estimates were shocking underestimates.

> *Following the invasion and conquering of islands in the Pacific it was evident that an invasion of Japan would be extremely costly. In the Battle for Okinawa alone allied forces lost close to 50,000 men. There were more than 90,000 Japanese military deaths and from 75,000 to 140,000 civilians either dead or missing. A total of more than a quarter of a million lives lost just to secure a foothold for invasion of the main islands.[2]*

Much like the battle for Okinawa, invasion of the home islands would be what Simon Bolivar Buckner, a Civil War Lieutenant General in the Confederate Army, called "Prairie Dog Warfare". Relatively unknown to soldiers in Europe and the Mediterranean it had been endured by soldiers and marines at Tarawa, Saipan, Iwo Jima, and Okinawa.

> *Prairie Dog Warfare was a battle for yards, feet, and sometimes inches. It was a brutal, deadly, and dangerous form of combat aimed at an underground, heavily fortified, non-retreating enemy.... at the early stage of the invasion, 1,000 Japanese and American soldiers would be dying every hour.[3]*

The Navy Insisted Invasion Was Not Required

Not surprisingly, the army and the navy had different ideas about how to end the war. The navy believed that a blockade, supported by an air campaign, would bring the end of the war without unnecessary spilling of blood. General McArthur did not believe that a blockage would force an unconditional surrender. Most top military leaders concurred with this line of thought: A blockade would not bring about quick defeat and while bombing might destroy cities, it would not destroy the Japanese army. It is not clear, if Japan were isolated, why they thought it urgent to destroy the Japanese army. Army officials argued that a blockade would take too long and the morale of the American public would suffer as a result. "They supported the use of an invasion that would go to the heart of Japan – Tokyo. The army got its way."4

The staff of Chester Nimitz estimated that 49,000 men would die in the first 30 days of the invasion of Kyushu. The army estimate was lower – General Marshall telling President Truman that 31,000 would die in that period. The Joint Chiefs of Staff estimated that the invasion of Kyushu would cost 109,000 lives with another 347,000 wounded. The invasion of both the home islands might cost 1.2 million casualties with 267,000 deaths. General Charles Willoughby, Chief of intelligence for General MacArthur, predicted that one million American men would be killed or wounded in the first year. Willoughby's own intelligence staff thought this estimate too low. The navy estimated there would be between 1.7 and 4 million American casualties with 400,000 t0 800,000 of these being deaths, and estimated that there would be up to 10,000,000 Japanese casualties.

We might wonder why invasion advocates could argue that a blockade, even a lengthy one, with minimal American loss of life would damage morale on the home front more than a beachhead assault killing tens of thousands of soldiers and marines. Were they simply planning to do what they were accustomed to doing – sending troops to invade another island and suffer horrific casualties? Although the beaches at Normandy presented terrible scenes of slaughter, as had the island campaign, an invasion of the Japanese mainland promised death and destruction on an unprecedented scale. "The battle for Japan might well have resulted in the biggest blood-bath in the history of modern warfare." 5

The apparent willingness of top level officials to accept a huge number of casualties to gain a quick defeat might cause one to wonder if their mind-set was reminiscent of Napoleon's boast to Metternich at Schonbrunn or Field Marshall Foch's repeated dispatch of French troops "over the top" in World War I.

The U.S. Invasion Plan

The first significant invasion objective was to be the island of Kyushu, which lay immediately to the south of the main island, Honshu. In what was called Operation Olympic, the invasion was to start with a series of landings on little islands west of Kyushu in October 1945 where planners thought U.S. troops could set up set up a radar installation to provide early warning for the approaching fleet, direct combat planes from the carriers, and establish a seaplane base. Troops would then pour ashore on Kyushu. In view of Japanese defense plans the expectation that troops could quickly set up a radar station and seaplane base was surely optimistic.

Admiral William Halsey's fleet, the third, would send planes to attack targets on the main island of Honshu. Admiral Raymond Spruance's force, the 3,000 ship fifth fleet, would follow with the invasion troops. Battleships, heavy cruisers, and destroyers would sit off the coast, shelling the beach and coastal areas with tons of high explosives. The invasion would begin early morning on the first of November. Thousands of marines and soldiers would be ordered to storm the beaches on the eastern, southern, and western coasts of Kyushu. If all went well, the invasion of Honshu, the main island of the Japanese homeland – called Operation Coronet – would take place the following spring. Coronet would be twice the size of Olympic, landing as many as 28 divisions on Honshu. With each division at 10,000 to 15,000 men that would be between a quarter and almost a half million men trying to fight their way ashore.

Few beaches were suitable for landing so the Japanese could anticipate where the troops would land. Planners back in the states decided that if additional troops were needed to replace losses, which they thought would be the case, other divisions shipped home from Europe to the U.S. for training would be shipped to Japan for the final push.[6]

Japanese Plans

Without being too harsh we might recall the cruel joke that an example of an oxymoron is the phrase 'army intelligence'. In the course of planning for the invasion, Allied Intelligence estimated the Japanese had only 2,500 aircraft available to resist the invasion force. They predicted that 300 would be used in suicide attacks. Unknown to U.S. military leaders, the Japanese had been keeping in reserve all aircraft, fuel, and pilots and had put an all out effort into building new planes. In August 1945, they still had 5,651 army and 7,074 navy – a total of 12,725 planes of all types. In addition, hidden in mines, tunnels, and even in the basements of department stores, more planes were being built. They had 20 suicide takeoff strips in southern Kyushu complete with underground hangers, 35 camouflaged airfields, and 58 more airfields in Korea, western Honshu, and Shikoku. The Japanese military hierarchy was determined to inflict such damage on the invading force that the Americans would back off, accept less than unconditional surrender, and enable them to save face.[7]

Kamikaze Attacks

At Okinawa alone, Japanese suicide attacks had sunk 32 Allied ships and damaged more than 400 others. But since American bombers and fighters daily flew unmolested over Japan during the summer of 1945, planners concluded that the Japanese had spent their air force and expected no more than 300 Kamikaze attacks.

Reality was far different. As the invasion fleet would approach on the night before the invasion the Japanese planned to launch a suicide attack against the fleet with 50 Japanese seaplane bombers, 100 former carrier aircraft, and 50 land based army planes. On invasion morning, while 2,000 army and navy fighters were fighting to the death to control the sky over Kyushu, a force of 330 navy combat aircraft would attack the main body of the task force to keep it from using its aircraft to protect the troop transports. While American fleet aircraft were thus engaged, another force of 825 suicide planes would hit the American transports. As the invasion fleet approached the beaches, they planned to launch another 2,000 suicide planes in waves of 200-300 aircraft.[8] How many troop ships loaded with soldiers and marines would have gone to the bottom of the Pacific is unknown.

Japanese planners knew that by midmorning, U.S. land based aircraft would have to return to base. Only shipboard gunners and fatigued carrier pilots, landing repeatedly for fuel and ammunition, would be left to defend the vulnerable troop transports. Both gunners and pilots would be exhausted but the waves of suicide attacks would keep coming – the Japanese would continue that onslaught for ten days while the invasion fleet sat off the coast attempting to land troops.

In addition the Japanese had 40 submarines left, some armed with long range (20 miles) torpedoes. The ships would also face suicide attack by midget submarines, human torpedoes, and exploding motorboats.

For Troops That Made It Ashore The Beaches Would Be Deathtraps

On shore the opposition would be fierce. As Americans soldiers and marines waded ashore they would encounter anti-landing obstacles and batteries of coast artillery. Those that survived to the beach would face a hail of shells from artillery, mortars, a heavy network of pillboxes, bunkers, and underground fortresses. They would have to work their way through barbed-wire entanglements set up to funnel them into the muzzles of Japanese guns.

In the bitter island-hopping campaign allied troops had outnumbered the dug in defenders by two to one or even three to one. Many of the fanatical defenders were poorly trained and poorly equipped but because they were dug in the islands had taken a huge toll of dead and wounded American young men. This would be worse. American marines and soldiers would face the well trained, well-fed home army with huge stockpiles of arms and ammunition – many the elite of the army, possessed with fanatical fighting spirit.

> *On the beaches and beyond would be hundreds of Japanese machine gun positions, beach mines, booby traps, trip-wire mines, and sniper units. Suicide units concealed in "spider holes" would engage the troops as they passed nearby. In the heat of battle, Japanese infiltration units would be sent to wreak havoc in the American lines by cutting phone and communication lines. Some of the Japanese troops would be in American uniform, English-speaking Japanese officers were assigned to break in on American radio traffic ... to confuse troops. Other infiltration units with demolition charges strapped on their chests or backs would attempt to blow up American tanks, artillery pieces, and ammunition stores as they were unloaded ashore.[9]*

Japanese Civilians Would Fight to the Death.

The code of military authorities was Bushido, the Way of the Warrior. They would encourage or force civilians to hold out to the death or even commit mass suicide to avoid 'losing face'. Shosango and Ketsugo war policies encouraged every single man, woman, and child to fight to the death. Japanese schools were closed and men, women, and children were trained to use any weapons available – from lunge mines to swords, axes, and bamboo spears – to kill Americans, often in night attacks. (One source consulted contained pictures of Japanese women lined up on the street, being trained in bamboo spear combat by a military instructor.).[10]

A technically educated American, part of a group interrogating a well-informed and intelligent Japanese army officer soon after the peace treaty was signed, provided this account.

> *We asked him what, in his opinion, would have been the next major move if the war had continued. He replied: "You would probably have tried to invade our homeland with a landing operation on Kyushu about November 1. I think the attack would have been made on such and such beaches."*
>
> *"Could you have repelled this landing?" we asked, and he answered: "It would have been a very desperate fight, but I do not think we could have stopped you."*
>
> *"What would have happened then?" we asked.*
>
> *He replied: "We would have kept on fighting until all Japanese were killed, but we would not have been defeated," by which he meant that they would not have been disgraced by surrender.*[11]

What Did Use of This Weapon Achieve?

Although it may have served as an example of the death and destruction it could cause and thus prevented its use during the cold war, this question can definitely be answered by looking realistically at the ultimate saving in the life of military troops on both sides and also Japanese civilians. Henry Miller refers to a study performed by physicist William Shockley for Secretary of War Henry Stimson's staff. Shockley estimated that the invasion of Japan's home islands would have cost 1.7 to 4 million American casualties, including 400,000 to 800,000 dead soldiers and marines and five to ten million dead Japanese civilians.

> *In the battle for Okinawa 75,000 to 140,000 civilians were dead or missing (the higher number being approximately the same as the total civilian deaths from both the Hiroshima and Nagasaki bombs). In 1945 Japan had a population of approximately 52 million, one hundred times as many as Okinawa. That would suggest the possible loss of somewhere around 17,000,000 civilian lives.*[12]

Many civilians in Japanese-controlled territory far from the home islands owed their lives to President Truman's decision to use the bomb. They were dying each day that went by from starvation, hard labor with little food, and harsh treatment. Put into camps the men were subjected to brutal slave labor. The occupying Japanese troops displayed Bushido disrespect for men who had not had the decency to kill themselves rather than to surrender. They were tortured regularly. The women also suffered. One of the sources consulted quoted an individual whose mother lived in occupied territory and furnished this account.

> *They were periodically subjected to group punishments. The one that lives in my mother's memory more than sixty years after the fact was the requirement that they stand in the camp compound, in the sun, for 24 hours. No food, no water, no shade, no sitting down, no restroom breaks (and many of the women were liquid with dysentery and other intestinal diseases). For 24 hours, they'd just stand there, in the humid, 90+ degree temperature, under the blazing tropical sun. ...The older women, the children and the sick died where they stood.*

> *Rations that started out slender were practically nonexistent by war's end. Eventually, the women in the camp were competing with the pigs for food. ... By war's end, my mother, who was then 5'2", weighed 65 pounds.... She had started to die. Had the atomic bomb not dropped when it did, my mother would have starved to death.*[13]

Foreign Affairs magazine estimated that in every month that the war continued in 1945, Japanese forces were causing the deaths of between 100,000 and 250,000 civilians in occupied territory.

American young men had already played a horrible price in the European theater and then in the grueling island hopping campaign in the pacific.

> *After four long years of war, American deaths were already about 292,000. In other words, the invasion of Japan could have resulted in the death of twice as many Americans as had already been killed in the European, North African, and Pacific Theaters!*[14]

As it was, "The aircraft carriers, cruisers, and transport ships scheduled to carry the invasion troops to Japan, ferried home American Troops in a gigantic operation called Magic Carpet."[15]

An Ironic Historical Note

There is another reason that the timing of the bomb's use was significant. Had the invasion gone ahead, the American fleets would have been positioned at Okinawa in October. In the words of Karl Compton, writing in *The Atlantic* in 1946:

> *If the atomic bomb had not been used, evidence like that I have cited points to the practical certainty that there would have been many more months of death and destruction on an enormous scale. Also the early timing of its use was fortunate for a reason which could not have been anticipated. If the invasion plans had proceeded as scheduled, October, 1945, Airplanes would have covered Okinawa and its harbors would have been crowded with landing craft poised for the attack. "The typhoon which struck Okinawa in that month would have wrecked the invasion plans with a military disaster comparable to Pearl Harbor."*[16]

Superior Technology Has Brought the U.S. Safety, What Might Inferior Technology Bring?

In the quotation that precedes Chapter 12 Stefan Possony and J. E. Pournelle tell us that other things being equal, battles are won by superior technology, but clearly superior technology prevents battles. The nuclear bomb was clearly superior technology. Its use prevented a battle that would have cost hundreds of thousands of American men their lives and millions of Japanese civilians theirs.

In the immediate post war years, when the Soviet Union threatened to invade Europe with overwhelming conventional forces, the threat of this weapon probably prevented conventional war. The U.S. extended the protection of its "Nuclear Umbrella" to Japan and allies in Europe. After the Soviets built their own force of nuclear weapons an uneasy balance of power developed. Although the U.S. chose to become involved in "limited wars" with Soviet and Chinese proxy states, the U. S. doctrine of Mutual Assured Destruction appeared to prevent all out war.

Recent years have seen constrictions on U.S. spending for advanced strategic weapons. Research is costly, fielding new systems is costly, and maintaining them once they are in the field is costly. America is preoccupied with economic problems. At the same time "little" wars – currently combat in the Middle East – drain defense coffers.

Piecing together numbers from various public sources, assuming the Army would spend none of its Research and Development budget on strategic defense research, the Navy half, and the Air Force three fourths (possibly very generous estimates) that would total a bit less than 24 billion dollars which seems like (and is) a great deal of money. However it works out to barely over one tenth of one percent (0.00133) of a U.S. GDP of approximately 18 trillion dollars.

The Trap of Being Number One

History teaches us that a nation that develops a superior technology for warfare finds itself in a superior position. It teaches also that there is an inevitable tendency for that nation to relax while a busy opponent outflanks it with something better. That 'something better' could be better technology, vastly greater numbers of equivalent systems, or both. The objective fact is that numbers of U.S. strategic systems are fixed or declining, and the doctrine for their employment, Mutual Assured Destruction, is now seventy years old.

The dark starting point for strategy must not be that which is possible: we must discover what is necessary and try to achieve it.

General d'Armee Andre Beaufe

But there is at least a minority of human beings – mostly male –who enjoy killing. That minority may be small, but it does not take many enthusiastic killers to trigger the destruction of a fragile society.

Ralph Peters

APPENDIX

A WAR AGAINST TERROR

The idea of going to war against a word may have had its beginning when Tom Brokaw, watching a video of one of the twin towers falling on 9/11 told viewers that "Terrorists have declared war on America." Since then the exact title of the activity the U.S. is engaged in seems to have evolved through a series of rethinking exercises. Five days later, speaking from Camp David, President Bush said "This crusade – this war on terrorism – is going to take a while." He later apologized for use of the term crusade, since critics suggested that some of those of the Muslim faith might be offended. On the 20th, he called it a war on terror which would begin with al-Qaeda, would not end "… until every terrorist group of global reach has been found, stopped, and defeated." [1]

The British government later decided that 9/11 was a crime and not an act of war. So what it was really doing was law enforcement. In a speech at Oxford University in 2013, Jeh Johnson, the General Counsel of the Department of Defence predicted that al Qaeda will be so weakened to be ineffective, and has been "effectively destroyed." [2]

Who had it right? Or did anyone? It would seem that President Bush came the closest.

When a people or a nation attempts to kill or nullify the power of people who first attempted to kill or destroy them that is surely a defensive war. An international campaign in the Middle East to stop takeover of Iraq and Syria by the Islamic State of Iraq and Syria goes beyond a law enforcement action. It is clear that real human beings in that area are taking territory, kidnapping, and beheading innocents – human with tanks, firearms, and artillery. It is a campaign against aggressors – not a war against a word.

Whatever we choose to call it, the United States and the rest of the civilized world now face a different kind of threat from what it has been accustomed to: a fanatical desire to torture and kill motivated by religious zeal. A radical element of one of the world's major religions, Islam, openly seeks to impose political rule on the rest of the human race. This radical element focuses on Jews, Christians, and members of that faith who do not subscribe to its radical views – and then on the rest of humanity. Members of this radical element claim it is a service to their God to subjugate, torture, or kill any outside its belief system.

This impulse to subjugate other peoples in the name of Islam is not new. Chapter four, in its discussion of weapon use by the Eastern Roman Empire, noted the fact that the Byzantines for centuries stood as the chief barrier to fanatical Arab armies that otherwise might have overrun Medieval Europe. At that time is was war between nations. Between the 7th and the 11th centuries the Byzantines fought back attempted conquest by expansionist Rashidun and Umayyad caliphs and their successors. Leaders of the Islamic State in 2016 seek to set up a caliphate in Syria, and open war rages there while a multitude of radical Islamic organizations and individuals have dedicated themselves to eradication of barriers to a world wide caliphate, imposing Sharia law on all.

The Longest Conflict in Human History

There is a common impression that Western civilization and Muslim military might first met during the Crusades, when, it is supposed, rapacious westerners descended on the Middle East in a frenzy of looting and pillage. This overlooks centuries of conflict prior to that time

Radical Muslim conquest started small but it soon turned into a continued war against the west. Mohammed led a raid for booty and plunder against a Meccan caravan in 1624, killing 70 Meccans. By the time of the death of Mohammed in 632, Muslims had conquered most of western Arabia and Southern Palestine. These were to a great degree "Holy wars". After 632 the new Muslim caliph, Abu Bakr, launched Islam into almost 1,500 years of continual bloody conquest and subjugation. [3]

By the tenth century the Eastern Roman Empire had lost territory in Syria and then recaptured it, but by 1030 the Turks gained power and began to push into Byzantine territory. The emperor of Byzantium, Alexius I Comnenus sent a delegation to Piacenza, Italy, asking Pope Urban II in March 1095 for help against the Turks. On November 27, 1095, in Clermont, France, Pope Urban II called for a crusade to help the Byzantines and to free the city of Jerusalem. Not surprisingly the history of that first Crusade is confused. The official start date was set as August 15, 1096. However a people's crusade left first in the form of five armies – three being defeated in Hungary and the other two at Nacaea in Northwest Turkey. Even the forces that left after the official start date were not one unified army.

Brought together in Constantinople 1096, they captured Nicaea and then Jerusalem in 1099. and defeated an Egyptian relief army that same year.[4]

The Huns and the Goths had previously invaded the Eastern Roman empire, coming from Western and Central Asia. One historian maintains that an invasion of the Byzantine empire in by the Avars from Northern China and Mongolia in 626 A.D. weakened the empire and allowed initial Muslim expansion.

Imperialistic and Colonialist Expansion of the Muslim Empire

Muslim wars of conquest have been directed for almost 1,500 years against hundreds of nations, over territories larger than the British Empire at its peak. Areas of conquest extended from southern France to the Philippines, to Austria, to Nigeria, to central Asia, to New Guinea. The Muslim goal was to have a central government. Damascus first, then at Baghdad, then at Cairo, or Istanbul, from whence local governors and judges were appointed by the central imperial authorities for far off colonies. Islamic law was forced on the local people, Arabic was introduced as the rulers' language, and heavier taxes were paid by the subjugated people than by their masters. In short, the Muslim conquerors set up a colonial empire. Even Muslims who were not Arab had less freedom that their colonial masters.

> *Christians and Jews could not bear arms -- Muslims could;*
> *Christians and Jews could not ride horses -- Muslims could;*
> *Christians and Jews had to get permission to build -- Muslims did not;*
> *Christians and Jews had to pay certain taxes which Muslims did not;*
> *Christians could not proselytize -- Muslims could;*
> *Christians and Jews had to bow to their Muslim masters when they paid their taxes; and Christians and Jews had to live under the law set forth in the Koran, not under either their own religious or secular law.* [5]

Muslim conquest of Christian North Africa appears to have been relatively easy until resisted by the Berbers and other native people west of Egypt. The North African people fought so strongly conquest in the west was brought to a near stop between Tripoli and Carthage for more than a quarter century. The Muslims finally broke through in a series of bloody battles which were followed by massacres of the largely Christian armies. Muslim conquest continued through North Africa, Spain, Portugal, and southern France, until stopped at an early battle of Poiters in the middle of France.

In the 700s, forcible conversion to Islam became Muslim policy.

> *There is a general view among the historians that the caliphs had begun to add a religious importance to their conquests, setting conversion to Islam as an important priority; their later caliphs had no first-hand remembrance of Mohammed; the vast distances of the empire led to independent rulers being established in Spain, North Africa, Cairo, Asia Minor, etc.; and the instability of the caliphates and resulting civil wars.* [6]

The history of the crusades does not paint a pretty picture, but, in regard to achieving a better understanding of the nature of the centuries-long war between radical Islam and other religions it is worthwhile to remember that by the time the first crusade was launched by Pope Urban II 1095, it had been preceded by centuries of Muslim wars of aggression. We should not deceive ourselves that these were peaceful expansions of dominion. For example, in 633 A.D. Muslim forces, led by Khalid al—Walid (called by Mohammad the Sword of Allah for his ferocity) conquered the city of Ullays along the Euphrates River (in today's Iraq). Khalid captured and beheaded so many that a nearby canal, into which the blood flowed, was called Blood Canal. A complete timeline of conquests can be found at the site referenced here.[7]

Current Trends in Terrorism

In the 1970s attacks aimed at killing people were limited and terrorists concentrated on material damage. The 1980s saw a move toward urban attacks and civilian targets. In the 1990s leftist organizations faded and groups driven by religious and religio-nationalist goals moved to the forefront. For example: in 1980 two out of 64 known groups were described as largely religious in their motivation; by 1995, the number was almost half, 26 out of 56, mostly guided by Islam.[8]

In the late 1960s, secular movements such as Al Fatah and the Popular Front for the Liberation of Palestine (PFLP0 started targeting civilians outside the immediate conflict area. After Israel's victory in the 1967 war, it became apparent that Arab countries could not defeat Israel. Using modern transportation and communications, radical Palestinians began a campaign of bombings, shootings, and kidnappings. The most notable of the latter was the kidnapping and death of Israeli athletes during the 1972 Olympic games at Munich. Key radical Palestinian groups included the Popular Front for the Liberation of Palestine (PFLP), the Popular Front for the Liberation of Palestine - General Command (PFLP-GC) which split off from the PFLP in 1968 and focused on terrorism, and the Abu Nidal Organization (ANO), which left the PLO in 1974 and has carried out terrorist attacks in 20 countries including the U.S., Great Britain, France, and moderate Arab countries. "By the end of the 1970s, the Palestinian secular network was a major channel for the spread of terrorist techniques worldwide."[9]

The Western World grew accustomed to reading and watching television reports of attacks on Israeli soldiers, and on Israeli civilians riding busses, shopping, or going to work. To many readers and viewers, that was something that happened "there" and not "here". To the extent it ever was however, the slaughter of men, women, and children by armed gunmen and suicide bombers is no longer confined to Israel or to Jews, a fact brought home to Europeans by attacks in Paris in November 2015 which left more than a hundred people dead and in Brussels, Belgium in March 2016, that saw deadly terrorist attacks at the city's primary international airport and a mid-city subway station. Then assassins struck in the United States. In December, 2015 a married couple shot and killed fourteen people in San Bernardino, California, an event investigated by the FBI as a terrorist act connected with ISIS (the Islamic State of Iraq and Syria).

Who are these people who use firearms on ordinary people going about their daily business and employ explosives to maim and kill? What should we call them? Politicians on the left side of the aisle, starting with the occupant of the White House in 2016 have refused to use the term "radical Islam". James Comey, however, Director of the FBI, "declared that 'radical Islamic terrorism' is an 'umbrella term' that accurately describes acts by 'savages ... (and) maniacs' in groups such as the Islamic State."[10]

Terrorist Groups

This list of organizations does not sum up the total terrorist threat but it provides a sample of the most notable organzations. In November, 2014, the *National Interest* described the first five as the five deadliest terrorist groups on the planet.[11]

The Islamic State of Iraq and Syria (ISIS)

This group is the first modern terrorist organization of today to seize territory and resources and use the resources of that territory to build an income stream to sustain its aggression. Treasury figures reported in 2014 placed its oil revenues at a million dollars per day. In mid-2015 its total revenues were placed at $80 million per month. However, by late April 2016 ISIS had lost one fifth of its captured territory to Kurdish, Syrian, and Iraqi forces. The number of people it ruled had fallen from nine million to six million. Tax revenue, about half of its funding, had fallen by approximately a fourth from its level in mid-2015.[12]

ISIS recruitment of volunteers for the battlefield is enhanced because it has done something no other terrorist organization has done: capture and hold territory. It has declared an Islamic caliphate in the center of the Arab world. ISIS carries out mass killings and executes anyone who shows any resistance. After the fall of the al-Tabqa air base in Syria it executed 250 Syrian troops. On another occasion it massacred over 200 Iraqi tribesmen. It is know most commonly by three names, the Islamic State of Iraq and Syria (ISIS), the Islamic State of Iraq and the Levant (ISIL), or simply the

Islamic State (IS). With an army in excess of thirty thousand men in 2014, it proved to be more than a match for largely ineffective Iraqi security forces. It took Kurdish *peshmerga* forces, Iran-backed militia forces, and U.S. airstrikes to recapture Amerli in 2014.[13]

The history of ISIS goes directly back to the Sunni terrorist organization al Qaeda – al Qaeda in Iraq (AQI). Following the U.S. invasion, that group, headed up by Abu Musab al-Zarqawi, carried out bombings and beheadings in Iraq. When Al-Zarqawi was killed in 2006 by an American airstrike, an experienced Iraqi fighter, Abu Du'a, took over the organization. Known also as AbuBakr al-Baghdadi, he had once been in U.S. custody in Iraq.[14]

AQI was seriously weakened in Iraq in 2007 as a result of what is known as the "Sunni Awakening" when the U.S. supported an alliance of Iraqi Sunni tribes fought against the jihadist group. However the Syrian civil war gave AQI new life and it moved into Syria. By 2013, al-Baghdadi had extended his group's influence back into Iraq, He changed the group's name to ISIS, "reflecting its greater regional ambitions," according to the U.S. State Department. Different translations of the Arabic name, as noted earlier, have given rise to other English-language names, including the Islamic State of Iraq and al-Sham (also ISIS) or the Islamic State of Iraq and the Levant (ISIL). Another name is Daesh, based on an Arabic acronym.

> *Although originally an al Qaeda affiliate, ISIS and al-Baghdadi had a public falling out in 2013 with Ayman al-Zawahiri, Osama bin Laden's replacement and leader of al Qaeda "core," over the role of another al Qaeda group, the* al-Nursa Front, *in Syria. In February 2014 a letter obtained by the* Long War Journal *reportedly showed al Qaeda's senior leadership was so fed up with al-Baghdadi that it severed all connection with ISIS. By declaring himself the "caliph" of the Islamic State in June 2014, al-Baghdadi appears to have challenged al-Zawahiri directly for the allegiance of all Muslim extremists. Ever since jihadist organizations around the globe have taken sides.[15]*

Some scholars believe that al-Qaeda and the Islamic State my soon merge. ISIS describes itself as the organization most faithful to Osama bin Laden's vision. Bitter rivalries with bin Laden's successor have continued, yet ISIS propaganda is reverential of bin Laden and respectful of al-Qaeda. Both organizations view the Western system of the nation state is inimical to the world wide imposition of Sharia law.[16]

Both the Islamic State and al-Qaeda welcome military intervention in Muslim lands. As the Islamic State proclaimed in 2014: "If you fight us, we become stronger and tougher. If you leave us alone, we grow and expand." [17]

The group has repeatedly called on its followers in Western nations to conduct deadly attacks in their home countries in retaliation for bombing of ISIS in Syria. A gunman in an attack in Paris in 2015 claimed to be part of ISIS, however other shooters had links to an al Qaeda organization. Shortly thereafter an Ohio man – a "self-radicalized" ISIS supporter was arrested and charged with plans to bomb the Capital building in Washington. Intelligence agencies in the West are concerned about men who travel to fight with ISIS in Iraq and Syria and gain combat experience, training in the use of weapons and explosives. When they return these fighters may be placed in contact with domestic terrorist networks that may be planning attacks.[18]

Did U.S. military action open the way for expansion of ISIS into Libya?

Although battered by events in Iraq in 2007, ISIS has since come back stronger than ever. Its success and slaughter in Syria has received a great deal of well deserved world attention, but this has not been its only notable success. It has established another significant beachhead. It is almost as if the United States and NATO had given ISIS Libya as a gift – as the saying goes – on a silver platter.

The administration of U.S. President Barack Obama had requested a resolution authorizing military intervention in Libya. "The goal, Obama explained, was to save the lives of peaceful, pro-democracy protesters who found themselves the target of a crackdown by Libyan dictator Muammar al-Qaddafi." [19] On March 17, 2011, the UN Security Council passed Resolution 1973. Two days later the United States and other NATO countries started bombing Qaddafi's forces. Seven months later rebel forces conquered the country and shot Qaddafi dead.

Later, the U.S. permanent representative to NATO, the supreme allied commander of Europe, and President Obama all declared that this had been a "model intervention."

> *That verdict, however, turns out to have been premature. In retrospect, Obama's intervention in Libya was an abject failure, ... Libya ...has*

devolved into a failed state. Violent deaths and other human rights abuses have increased several fold. Rather than helping the United States combat terrorism, as Qaddafi did during his last decade in power, Libya now serves as a safe haven for militias affiliated with both al Qaeda and the Islamic State of Iraq and al-Sham (ISIS). The Libya intervention has harmed other U.S. interests as well: undermining nuclear nonproliferation, chilling Russian cooperation at the UN, and fueling Syria's civil war.

Despite what defenders of the mission claim, there was a better policy available—not intervening at all, because peaceful Libyan civilians were not actually being targeted. Had the United States and its allies followed that course, they could have spared Libya from the resulting chaos and given it a chance of progress under Qaddafi's chosen successor: his relatively liberal, Western-educated son Saif al-Islam. Instead, Libya today is riddled with vicious militias and anti-American terrorists ...Immediately after taking power, the rebels perpetrated scores of reprisal killings, in addition to torturing, beating, and arbitrarily detaining thousands of suspected Qaddafi supporters. The rebels also expelled 30,000 mostly black residents from the town of Tawergha and burned or looted their homes and shops, on the grounds that some of them supposedly had been mercenaries.

Amnesty International issued a report last year (in 2014) that revealed ... savage mistreatment: "Detainees were subjected to prolonged beatings with plastic tubes, sticks, metal bars or cables. In some cases, they were subjected to electric shocks, suspended in contorted positions for hours, kept continuously blindfolded and shackled with their hands tied behind their backs or deprived of food and water." ... Ongoing attacks in western Libya, the report concluded, "amount to war crimes." As a consequence of such pervasive violence, the UN estimates that roughly 400,000 Libyans have fled their homes, a quarter of whom have left the country altogether.[20]

ISIS is said to have had approximately 6,000 fighters in Libya as of mid 2016. It also controls nearly 200 miles of Libyan coast line. The country serves as a safe haven from bombings in Syria and Iraq. It is moving to threaten Tunisia. Ominously, its occupation of Libya allows ISIS to be in closer proximity to Europe. This fact is coupled with information from the Iraqi government that nuclear material has been seized by insurgents there. [21] "Radiation poisoning and billions of dollars in financial losses would result if radio isotopes were used as a 'dirty bomb' and detonated in a major city."[22]

A long list of attacks by ISIS on the west can be found at the site referenced here.[23]

Boko Haram

Boko Haram, founded in 2001, is an organization not publicized in America but one that has been a plague in Nigeria. In Northern Nigeria, Boko Haram has destroyed entire villages and burned inhabitants alive or executed them with bullets to the head. The Council of Foreign Relations has estimated approximately 7,000 deaths. This estimate may be low. In a night raid on the border town of Chibok, militants captured 300 Nigerian schoolgirls and threatened them with forced conversion or forced marriage. This organization is suspected of being behind the killing of more than fifty schoolchildren by a suicide bomber. The Nigerian government has responded by declaring three northeastern states to be in a state of emergency, and the army itself has used brutal tactics against civilians, including executing of suspected Boko Harm sympathizers, many innocent.[24]

In April 2016, the commander of the United States military's Special Operations in Africa, Brig. Gen. Donald C. Bolduc, called attention to a weapons convoy intelligence sources judged to be from Islamic State territory in Libya headed for the Lake Chad region, an area where Boko Haram had wreaked havoc. American military officials say this is one of the first clear indicators of a direct link between the two extremist groups since Boko Haram pledged allegiance to the Islamic State last year – evidence that two of the world's most feared terrorist groups have begun to collaborate more closely. Such an event could indicate that they are working together to attack American allies in North and Central Africa in the future.[25]

Islamic Revolutionary Guard Corps – Quds Force

News stories about Iran have made the name of Iran's Revolutionary Guard familiar to many, but Quds Force is a different organization. (The Revolutionary Guard or 'Arm of the Guardians of the Islamic Revolution' is a branch of Iran's armed forces. While Iran's borders are guarded by the regular armed forces, the Revolutionary Guard is intended to protect the country's Islamic system.)

Quds Force is an external wing of the Revolutionary Guard. Engaged in covert activity, it is highly secretive. One journalist called its leader, Qassem Suleimani, "The Shadow Commander," Its activities appear to be instrumental in allowing Iran to exert influence – through money, arms, and battlefield advice – over a number of organizations: Lebanese Hezbollah, Kataib Hezbollah, Asaib Ahl al-Haq, Hamas, and the Palestinian Islamic Jihad.[26]

Haqqani Network

This is said to be the most deadly group operating in Iraq, largely because of its tribal relationships in Eastern Iraq and its cooperation with other terrorist groups. For example, at least five other organizations are said to rely on Haqqani Network for access across the Afghan-Pakistani border and for advice on operations against government security forces and coalition troops. These five are Al Qaeda, the Islamic Movement of Uzbekistan, the Pakistani Taliban, the Turkistan Islamic Party, and the Quetta Shura Taliban. Haqqani Network has existed for decades in that area, and will remain a threat to stability into the future.[27]

Kataib Hezbollah

This is a recently formed group, probably 2006 or 2007. It is not the same organization as Hezbollah based in Lebanon. Kataib Hezbollah, operating in Iraq, specialized in placing roadside IEDs on the roads used by U.S. Humbeees and ambushing U.S. troops conducting patrols of Iraqi neighborhoods. Ironically, with collapse of the Iraqi army, the Iraqi government has come to depend on Shia militias to reinforce its army fighting ISIS. These militias are believed responsible for imprisonment or murder of Sunnis under the slightest suspicion of sympathy toward ISIS. "If ISIL elicits shivers and fear in the heart of an Iraqi Shia, Kataib Hezbollah produces the same emotion in an Iraqi Sunni."[28]

Hezbollah

Hezbollah is a Shiite Muslim political party and militant group. Hezbollah (or "Party of God") was formed during Lebanon's fifteen-year-long civil war (1975-1990). Sometimes described as a "state within a state" it maintains a social services network in Lebanon. It was responsible for the suicide truck bombing of the U.S. Marine barracks in Lebanon in 1983 and the kidnapping of U.S. and other western hostages in Lebanon. Its reason for being is opposition to the State of Israel and involvement of Western Powers in the region, particularly the United States. As it became involved in the Syrian Civil War, fighting on the side of the Assad regime, it apparently suffered reprisals in Lebanon from rebels who are predominantly Sunni Muslim.[29]

The Muslim Brotherhood

The Muslim Brotherhood was founded in Egypt in 1928. It was said to combine political activism with Islamic charity work.

> *The Brotherhood's stated goal is to instill the* Qur'an *and* Sunnah *as the "sole reference point for ... ordering the life of the Muslim family, individual, community ... and state." Its mottos include "Believers are but Brothers", "Islam is the Solution", and "Allah is our objective; the Qur'an is the Constitution; the Prophet is our leader;* jihad *is our way; death for the sake of Allah is our wish". It is financed by members, who are required to allocate a portion of their income to the movement, and was for many years financed by Saudi Arabia, with whom it shared some enemies and some points of doctrine.[30]*

The so-called Arab Spring brought political power. The Muslim Brotherhood's candidate Mohammed Morsi was elected president of Egypt. His tenure was marked by widespread frustration with economic mismanagement and poor governance, and his administration was ousted by the military in July 2013. A violent crackdown followed in which Morsi, much of the Brotherhood's leadership, and thousands of its supporters were arrested.[31]

Saudi Arabia, a one time supporter, has declared the Muslim Brotherhood a terrorist organization as have Egypt, Syria, and the United Arab Emirates. Following a study that took a year and a half, the United Kingdom declared the Muslim Brotherhood a terrorist organization, describing it as "anti-democratic, openly supportive of terrorism, dedicated to establishing an Islamist government — and opposed to the rule of law." The Obama administration slammed the British document as flying in the face of the Brotherhood's history as a "nonviolent Islamist group." [32] According to one news report:

> *President Barack Obama has long embraced the Muslim Brotherhood and has defended and promoted Islam in an effort to fight Islamophobia. In January,* the State Department *hosted a delegation of leaders aligned with the organization. In 2011, Director of National Intelligence* James Clapper told Congress *that the Muslim Brotherhood has 'pursued social ends, a betterment of the political order in Egypt'.*[33]

Hamas

A spinoff of Egypt's Muslim Brotherhood, Hamas was founded in 1987. This was during the first intifada and Hamas eventually moved to the forefront of conflict with Israel. The United States and the European Union place Hamas in the category of terrorist organization. Its principal rival party, Fatah, dominates the Palestine Liberation Organization which has formally renounced violence. In 2006 Hamas candidates won Palestinian elections but the government was dismissed the next year. Fatah again took over authority of the West Bank and Hamas the Gaza Strip.

Although the border between Egypt and Gaza was officially closed in 2006-2007, a complex network of more than a thousand tunnels led from Egypt into Gaza through which flowed cash, construction material, and arms. Hamas levied a tax on the flow which provided a major source of revenue. Following the establishment of Abdel Fattah al-Sisi's military-backed government in Egypt, that country views Hamas as an extension of the Muslim Brotherhood, now outlawed in Egypt. As part of an anti-insurgency campaign the Egyptian army shut down the tunnels.

Some funding is said to come from Palestinian expatriates and wealthy individuals in the Gulf Area. Some Islamic charities in the West had sent money to Hamas-backed charities before the U.S. Treasury froze assets. Friction developed between Iran and Hamas due to the Iranian government's support for Bashar al-Assad in Syria.[34]

Al-Qaeda

After 9/11 al-Qaeda ("the base" in Arabic) became the most noted terrorist organization, more so even than Hamas or Hezbollah. It had its beginning during the Soviet occupation of Afghanistan when Middle Eastern men traveled to Afghanistan to join the fight. Osama bin Laden became the prime financier for an organization that recruited *mujahideen* from mosques around the world. Sometimes referred to as "Afghan Arabs" they numbered in the thousands. When the Soviets pulled out bin Laden returned to his native Saudi Arabia where he set up an organization to help veterans of the Afghan war.

For a time al-Qaeda and the name of Osama bin Laden became almost synonymous, however he did not run the organization alone. Dr. Ayman al-Zawahiri, who became his successor, was his number one advisor. Al-Zawahiri **was** an Egyptian surgeon from an upper-class family who served three years in an Egyptian prison on charges connected to the assassination of Anwar Sadat. Following his release he went to Afghanistan, where he began to work with bin Laden.

When Iraq invaded Kuwait in 1990 bin Laden became extremely angry because the Saudi government allowed U.S. troops to be stationed in Saudi Arabia. In 1991 the Saudi government expelled him. He established headquarters for al-Qaedi in Sudan, where he organized attacks on U.S. soldiers. In August 1996 he issued a Declaration of War against the U.S.

He organized an alliance of terrorist organizations he labeled the "International Islamic Front for Jihad Against the Jews and Crusaders". It included the Egyptian al-Gama'at al-Islamiyya, the Egyptian Islamic Jihad, the Harakat ul-Ansar, and others. When Sudan expelled him because of pressure from the U.S. and Saudi Arabia, he went to Afghanistan where he set up training camps and from where he directed the organization.

Unlike other terrorist organization al-Qaeda does not depend on sponsorship from a state such as Iran nor is it defined by a particular war. It is said to "operate like a franchise" [35] providing support and name recognition to terrorist groups operating in locations such as Algeria, Chechnya, Tajikistan, Somalia, Yemen, Kashmir, and the Philippines. Organizations that do not receive support may claim affiliation with al-Qaeda just for prestige.

Al-Qaeda's leadership oversees a loosely organized network of cells. It can recruit members from thousands of "Arab Afghan" veterans and radicals around the world. Its infrastructure is small, mobile, and decentralized—each cell operates independently with its members not knowing the identity of other cells. Local operatives rarely know anyone higher up in the organization's hierarchy. The principal stated aims of al-Qaeda are to drive Americans and American influence out of all Muslim nations, especially Saudi Arabia; destroy Israel; and topple pro-Western dictatorships around the Middle East. Bin Laden also said that he wishes to unite all Muslims and establish, by force if necessary, an Islamic nation adhering to the rule of the first Caliphs.

According to bin Laden's 1998 fatwa (religious decree), it is the duty of Muslims around the world to wage holy war on the U.S., American citizens, and Jews. Muslims who do not heed this call are declared apostates (people who have forsaken their faith).[36]

Al-Qaeda's ideology, "jihadism," includes a willingness to kill both "apostate", and Shiite Muslims.

Other Organizations

Other terrorist organizations exist as designated by the U.S. government. More detailed information can be found in the State Department Report, Country Reports on Terrorism, 2014.[37]

The Threat to the United States

Although their way may have been cleared in some instances by inept U.S. foreign policy, it is apparent that terrorist groups and networks have enjoyed sufficient resources to take advantage when an opportunity presented itself. Osmma bin Laden and al-Qaeda, Saudi Arabia and the Muslim Brotherhood, and Iran and Hesbollah. According to some authorities the 2015 nuclear treaty with Iran will open the way for Iran to increase its subsidies to terrorist organizations, along with its influence in the region and wherever the tentacles of these organizations may reach.

Since the treaty was signed, Iran has increased its financial support for Hamas and Hezbollah. Sanctions had beaten down Iran's economy, limiting that support but the government of Iran obviously believes that tens of billions of dollars – or more – will flow into the country in the coming years as a result of sanctions relief. This flow of money has enabled Hezbollah to purchase new armaments, including advanced technologies that many militaries around the world would envy. A Kuwaiti newspaper has reported that Hezbollah has obtained advanced weapons that Syria has received from the Russians.[38]

Terrorists Have Some Significant Advantages

Chapter Two pointed out the critical importance of range – that is, distance from the person or persons controlling a weapon and the point where it will impact its target. If time is not a critical factor – and it does not seem to be – terrorist groups actually have global range. With willing suicide bombers at the ready, a terrorist official can dispatch a human bomb to almost any point in the world and expect pinpoint accuracy. Since his objective is to destroy property – and particularly lives – by creating terror, he has something approaching a master weapon until some means is developed to intercept it. If he should choose to employ gunfire or chemical, biological, or nuclear material at the target he has a "smart weapon" that can perform the exact mission.

Any location where people are gathered closely together would seem to be a potential target. It has been said that Generals are always prepared to fight the last war rather than the next one. In a similar vein it could be said that the Transportation Security Administration has set up an elaborate defense system to protect airplanes. In doing so however it has herded people together like cattle in winding lines, making them a prime target.

Use of Modern Technology

Whatever their target – airports, shopping centers, schools – terrorist bosses have a highly lethal weapon at their disposal, if they can obtain an adequate supply of what we might call slow moving but long range missiles. To obtain that supply they have made good use of an advanced technology, and developed appropriate doctrine for its employment. In a recent interview a security expert described how ISIS, for one, has made adept use of social media.

> *ISIS ... is radicalizing American citizens who are on their computers sitting in their bedrooms or downloading ISIS propaganda on their smart phones. And these Americans no longer have to travel to some overseas training camp to be indoctrinated by ISIS – the group is also specifically catering to English-speaking militants and encouraging them to carry out attacks in countries such as the United States.[39]*

ISIS publishes instructions in English on how to communicate in encrypted mode and tells its recruits to use something called the Tor browser that will disguise the IP address of the user. It released a booklet providing tips on making bombs, hiding weapons, and avoiding surveillance by the police – an instruction book on how to be a 'secret agent' in a non-Muslim country.[40]

Modern technology has provided fearsome methods to kill, for terrorist groups to kill large numbers of people, including :Americans – possibly home grown terrorists seduced via the internet. Al-Qaeda for example

> *Had a focused nuclear weapons program and repeatedly attempted to buy stolen nuclear bomb material and recruit nuclear expertise. It went as far as carrying out crude tests of conventional explosives for their nuclear bomb program in the Afghan desert.[41]*

ISIS apparently has carried that quest farther. The Islamic State

> *Has more money, controls more territory and people, and enjoys a greater ability to recruit experts globally than al-Qaeda at its strongest ever had. The group's apocalyptic rhetoric, envisioning a final war between itself and the 'crusader' forces, suggests a need for very powerful weapons, and recent incidents such as the in-depth monitoring of a senior official of a Belgian facility with substantial stocks of HEU (highly enriched uranium) are worrying indicators of possible nuclear intent.[42]*

Contrary to popular thought, the nuclear threat is not limited a nuclear 'bomb' as such. There are three types of nuclear or radiological terrorist attack. Although terrorists might get possession of or simply detonate a weapon made by a national state, it is conceivable that it could make a crude nuclear bomb from stolen separated plutonium. Nuclear sabotage is the second, causing a meltdown similar to the accident at Fukushima, or sabotaging a spent fuel pool or a high-level waste storage site. A third ominous possibility is the assembly of a "dirty bomb" or other uses of radiological materials which might be stolen from hospitals, industrial sites or other sources.

> *A far simpler approach (versus a full scale nuclear 'bomb') would be for terrorists to obtain radiological materials ... and disburse them to contaminate an area with radioactivity, using explosives or any number of other means. In most scenarios of such attacks, few people would die from the radiation – but the attack could spread fear, force the evacuation of many blocks of a major city, and inflict billions of dollars in costs of cleanup and economic disruption.*[43]

Use of The Internet to Recruit Domestic Terrorists

The Director of the FBI has expressed concern about the use of social media by terrorist organizations.

> *It's a message that goes to troubled souls anywhere in the United States. As a result, we have investigations in every state in the Union that focus on trying to understand who is consuming that poison.*[44]

Terrorists use powerful tools – online message boards and chat rooms – to communicate, coordinate attacks, spread propaganda, raise funds, and recruit individuals to their cause. They can offer tutorials on building bombs, shooting American soldiers, and firing surface-to-air missiles. Pentagon analysts have indicated that they monitor as many as five thousand web sites, concentrating on fewer than a hundred. Hoax sites make it difficult to track them. Interesting enough, in Iraq, in 2009, terrorist propaganda sites were watched by large segments of society, and videos were apparently "sold in Baghdad video shops, hidden behind the counter along with pornography."[45]

As early as 2007, when it was estimated that there were as many as 5,000 jehad web sites, individuals in the front line of 'the war on terror' as it was called, recognized the power and reach of the internet. When interviewed by Scott Pelly of "Sixty Minutes", Army Brigadier General John Custert who was head of intelligence at central command, responsible for Iraq and Afghanistan, summed up the power of the internet to radicalize Islamic youth and create "an inexhaustible supply of suicide warriors":

You start off with a site that looks like current news in Iraq. With a single click, you're at a active jihad attack site. The real meat of the jihad Web site, Jihad Internet. Beheadings, bombings, and blood. You can see humvees blown up. You can see American bodies drug through the street. You can see small arms attacks. Anything you might want in an attack video. Next link will take you to a motivational site, where mortar operatives, suicide bombers, are pictured in heaven. You can you see their farewell speeches. Another click and you're at a site where you can download scripted talking points that validate you have religious justification for mass murder. ... Without a doubt, the Internet is the single most important venue for the radicalization of Islamic youth.

That's what Abu Musab al Zarqawi understood when he used the Internet to promote himself as head of al Qaeda in Iraq. When Abu Musab al Zarqawi, a street thug, beheaded an American businessman he became a rock star over night....The same with American contractor's bodies in Fallujah in April of '04. The same is Daniel Pearl having his throat slit having a gun put to his head. The power of the Internet is unbelievable.[46]

As internet usage has increased, terrorist groups have become increasingly able to manipulate the grievances of alienated young people. At the same time there is a growing trend of self-radicalization. In the latter case young people use the internet to acquaint themselves with the ideologies of terrorist groups. Social networking sites and instant messaging programs allow constant communication. In its anti-Israli recruiting Hamas targeted young children on its al-Fateh (the conqueror) site with children's stories and Disney-like characters. Hizballah made available first-person shooter games called "Special Force and Special Force 2" the latter of which is available in Arabic, Farsi, and English. These were directed at Israeli soldiers but the application to American targets is exemplified by another on-line game "Night of Bush Capturing" where the objective of the game is for players to hunt and kill former

President George W. Bush.[47]

Opinion as to the severity of the threat varies widely among individuals who might be expected to know. The chief of homeland security in 2015 stated that "we have no specific credible intelligence about a threat of the Paris-type directed at the homeland here" but did note that "We are always concerned about potential copycat acts, home-born, home grown violent extremism ..."[48]

More Radicalized Muslims Means More Terrorist Attacks

Omitting 9/11, if we compare the severity and scale of attacks in Europe with those suffered by the United States, there may be a reason why Europe has seen the brunt of attacks in 2015-2016: the number of individuals who have left their homes to go to the Middle East to engage in their Holy War and then returned. The Belgian government knows that approximately 200 Belgians are fighting in Syria and could return to their homeland and carry out terrorist attacks. Since the attack on the Brussels airport and a subway station that injured more than 300 people and took the lives of 32, police there are attempting to discover more about the Islamic State cell they believe responsible. At least 6,900 militants from Western countries have traveled to Syria since 2013. Few have been Americans however. Seven of these have returned. One went back to Syria to carry out a suicide attack. One has been charged with a plot to kill American soldiers at a base in Texas.[49]

Starting in the fall of 2014 there have been five attacks in the U.S. that officials attribute to individuals inspired by ISIS. In the fall of that year a man attacked police in New York with a hatchet. In May of the following year two men opened fire in Garland, Texas at a cartoon contest of the Prophet Mohammed. It was found that they had been in direct contact with members of ISIS. Approximately six months later a student at the University of California, Merced stabbed four people. He has been on ISIS web sites. In January 2016 a man who shot a policeman in Philadelphia was quoted as saying "I pledge my allegiance to the Islamic State, and that's why I did what I did."[50]

One analysis looked at how previous waves of American volunteers had fared as they tried to make their way to and from jihadist camps in places like Somalia or Afghanistan. Most were native-born or naturalized U.S. citizens and either Muslims by birth or zealous converts. Some were what might be called "jihadi tourists' "snapping selfies" and returning to brag to friends. A third had been arrested before they could carry out their plans, usually on charges of attempting to provide material support to a terrorist organization. A third were arrested or killed overseas and another third made it back. Most of those were arrested. Roughly one in ten slipped back into society.

In Europe, the threat is much more significant. There, returning fighters number in the thousands, not dozens. "And it may worsen in the coming years as a new generation of fighters—drawn by online images of beheadings and bloodshed—makes its way home from the fighting in Syria and Iraq." [51]

How Do We Defend Ourselves?

What does the defense need? New technology or better employment strategy for technology that already exists? Or both? To find the best answers we might begin with words penned by Ralph Peters as the 20th century was drawing to a close.

> *The United States is unbeatable on a traditional battlefield, but that battlefield is of declining relevance.... Beware of any enemy motivated by supernatural convictions or great moral schemes. Even when he is less skilled and ill-equipped, his fervor may simply wear you down. Our military posture could not be more skewed. We build two-billion-dollar bombers, but we cannot cope with bare-handed belief.* [52]

News stories have told us that insurgents in Iraq have used cell phones to trigger Improvised Explosive Devices under roads traveled by Humvees carrying American soldiers. Terrorist organizations have made extensive use of social media to radicalize and recruit susceptible Muslims – particularly young men – to their cause.

Drones have become ubiquitous. Authorities are properly concerned that they might somehow be put to nefarious use. As a defense against known terrorists they could perhaps be used with light weight cameras and face recognition technology to screen crowds. However the implications of invasion of privacy in a free society cannot be lightly disregarded.

High speed, high end computers could sift through emails looking for suspicious messages with terrorist implications unknown to the population thus being watched –

in fact they may already be. However, here again, the possibilities for abuse by government are huge and quite possibly (and in the view of this writer) not worth that cost.

In an age when electronic data transfer and storage is vital to the flow of commerce – hacking or sabotage of computer systems is a threat with potentially huge consequences. The "Y-2 K" threat may not have been as serious as visualized in the 1990s, or it may have been, and the sometimes-frantic fixes, patches, and workarounds that were developed to combat that perceived threat may have prevented serious consequences. The consequences of terrorist actions to our computer systems may be at least as serious as any visualized from a century shift then. This would seem to be an area where defenders of computer systems need to stay ahead of potential hackers.

The economic and human cost of an EMP device, a dirty bomb, or even a conventional terrorist attack in an urban area or attempted destruction of major railroad or highway bridges or the power grid could easily climb into the billions. The possible human toll is incalculable. An adequate defense is worth whatever the dollar cost.

It has been said that our system of policing works because almost all people are basically honest. As the number of destructive agents climbs there are not enough policemen to go around. Individual citizens must fill the breach, or order will perish.

Two generations ago, if a farmer saw a large rock or a fallen branch on a rural road, he would stop and remove it. In the unlikely event someone else came by, he most likely would stop to help. Today, on a busy freeway, the person who first stopped would likely be run over and no one would stop to help. It is a job universally left to the professionals. Perhaps to some degree, our society would benefit from a return to the philosophy that safety is the responsibility of all of us. For example, in every state that has instituted legal concealed carry of firearms violent crime rates have plunged – in particular, crimes against the weak, females and the aged. This is simply because a potential perpetrator never knows who might be armed and who might not and if they guess wrong, a trained, armed citizen, if present, may prevent serious harm to the innocent.

Is it possible that in a troubled future, citizens of a free society may have to bear more of the burden of protecting vital nodes of our civilization? Of primary importance might be accepting the responsibility to inform authorities immediately when something appears out of order, and secondly, if they have the requisite will and training (and clearly not all would be willing to undergo such training) being prepared to be the first line of defense against violent terrorism in any form.

Perhaps some innovative technology is out there to turn the tables against terrorists and their masters. Perhaps there is not. If there is not, then better use of the technologies we have will be required. With no obvious "new thing" in technology seemingly available, part of the solution may be a change in philosophy – in methods of employing what is available to us. Unfortunately, there is always a resistance to change. Would American citizens today be ready to step up and assume at least some of the responsibility for their own safety and that of their neighbors?

The starting point to solution of the seemingly unsolvable terrorist problem may be the quotation that leads off this discussion:

> *"The dark starting point for strategy must not be that which is possible: we must discover what is necessary and try to achieve it.*

Reference Notes

Chapter 1 The Continued Existence of War

1. Palit, D. K. *War in the Deterrent Age*. (New York: A. S. Barnes and company, 1966). PP, 89 – 90.

Chapter 2 Technology Makes Useful Weapons, Appropriate Doctrine Makes Those Weapons Useful

1. Fuller, J. C. F. *Armament and History*. (New York: Charles Scribner's sons, 1945). PP. 7 – 8

2. Fuller, P.8.

3. "Russia fires intercontinental ballistic missile in scheduled test" n.p. n.d. <http://rt.com/news/russia-test-fire-icbm-825/>

4. Gutterman, Steve. "Russia plans new ICBM to replace Cold War 'Satan' missile by 2018". Reuters. n.d. <http://www.reuters.com/article/2013/12/17/us-russia-missiles-idUSBRE9BG0SH20131217>

Chapter 3 Warfare: Sometimes Limited, Sometimes Unlimited

1. Preston, Richard and Sidney wise, *Men in Arms*. (New York: Praeger publishers. 1970) PP.16 – 19.

2. Creasy, Sir Edward . *15 decisive battles of the world*. (written in 1851) (Harrisburg, Pennsylvania: the telegraph press, 1960). P.2.

3. Preston and wise.

4. Fuller. *Armament and History*, P.55.

5. Fuller, *Armament and History*, P.54.

6. Fuller, *Armament and History*, P.96.

7. Preston and wise, PP.184, 192.

8. Fuller, *Armament and History*, PP.125 – 126

Chapter 4 The Search for Range: Early Weapons

1. Cleator, P. E. *Weapons of War*. (New York: Thomas P. Crowell well company, 1968), P.38.

2. Bernie, Arthur. *The Act of War*. (New York: Thomas Nelson and sons, limited, 1942), P.13.

3. Preston and Wise, P.21.

4. Bernie, P.20.

5. Bernie, P.25.

6. Bernie, P.27.

7. Bernie, PP.34 – 35.

8. Preston and Wise, P.51.

9. Fuller, J. C. F. Decisive Battles: *Their Influence Upon History and Civilization.* (New York: Charles Scribner's sons, 1940), P.206.

10. Preston and Wise, P.60.

11. Preston and Wise, PP.56 – 57.

12. Preston and Wise, PP.56 – 57.

13. Preston and Wise, PP.52 – 53.

14. Fuller, *Armament and History*, P.60.

15. Fuller, *Armament and History*, P.60 – 61.

16. Konstam, Angus. *Byzantine Warship vs Arab Warship*. (2015 Osprey Publishing) pp. 5. 72-74

17. Fuller, *Armament and History*, P.62.

19. Konstam. pp. 38-40)

20. Preston and Wise. P.61.

21. Cleator, P.58.

22. Fuller, *Armament and History*. P.60.

23. Cleator. PP.105 - 106.

24. Cleator. P.110.

25. Fuller, *Armament and History*. P.71.

26. Preston and wise, PP.89 - 89; Fuller, *Armament and History*. PP.70 - 72; Cleator. P.110.

27. Preston and Wise, P.90.

Chapter 5 The Search for Range The Age of Gunpowder

1. Cleator. PP.104 - 105.

2. Fuller. *Armament and History*, P.79.

3. Cleator, PP.122 - 127.

4. Cleator, PP.132, 136.

5. Cleator, PP.123 - 124.

6. Cleator. P.127.

7. Cleator. P.128.

8. Cleator. P.130.

Reference Notes

9. Fuller, *Armament and History*, P.83.

10. Cleator, P.124.

11. Fuller, *Armament and History.* P.82.

12. Preston and Wise, P.98.

13. Fuller. *Armament and History*, P.83.

14. Preston and wise, P.103.

15. Fuller, *Armament and History*, P.84.

16. Fuller, *Armament and History,* PP.97 – 98.

17. Cleator. P.164.

18. Fuller, *Armament and History*, P.118.

19. Cleator, PP.164 – 165.

20. Fuller. *Armament and History*, PP.120 – 121.

21. Preston and Wise, P.107.

22. Fuller, *Armament and History*, P.86.

23. Cleator. P.135.

24. Cleator. PP.134 – 135.

25. Cleator. P.139.

26. Cleator, PP.140 – 141.

27. Fuller, *Armament and History*, P.118.

28. Cleator. P.138.

29. Fuller. *Armament and History*, PP.110 – 111.

30. Cleator. P.153.

31. Cleator. P.155.

32. Cleator. PP.158 – 159.

33. Cleator. P.160.

34. Cleator. P.174.

35. Palit. PP.36 – 48.

Chapter 6 The Search for a Better Shield

1. Cleator. P.53.

2. Cleator. P.54.

3. Cleator. P.55.

4. Cleator. P.61.

5. Cleator. P.75.

6. Cleator. PP.97 – 98.

7. Cleator. P.97.

8. Fuller. *Armament and History*, P.57.

9. Cleator. P.102.

10. Palit. P.47.

11. Palit, P.85.

12. Palit. P.86.

13. Cleator. P.71.

14. Creasy. P.26.

15. Cleator. PP.111 – 112.

Reference Notes

16. Cleator, P.75.

17. Palit. PP.47 – 48, 58.

18. Cleator. PP.103 – 104.

Chapter 7 The Quest for Mobility

1. Fuller, *Armament and History*, P.35; Cleator, P.73.

2. Fuller, *Armament and History*, P.33.

3. Fuller, *Armament and History*, P.33.

4. Fuller, *Armament and History*, P.33.

5. Cleator. PP.47 – 48.

6. Creasy. P.78.

7. Cleator. PP.50 – 51.

8. Creasy. P.78.

9. Cleator. P.95.

10. Preston and Wise, PP.28 – 30.

11. Cleator, P.83; Fuller, *Armament and History*, P.17; Preston and wise, P.67.

12. Cleator. P.95.

13. Cleator. PP.60 – 61.

Chapter 8 Sail and Gunpowder

1. Fuller, *Armament and History*, PP.88 – 89.

2. Fuller, *Armament and History*, PP.90 – 91.

3. Creasy. PP.133 – 135.

4. Creasy. PP.159 – 160.

5. Creasy. P.248.

6. Creasy. PP. to 248 – 249.

7. Creasy. PP.233 – 310.

8. Creasy. PP. 250–251.

9. Creasy, P.253.

10. Creasy. P.254.

11. Creasy. PP.257 – 258.

12. Creasy. P.266.

13. Creasy, P.269.

14. Creasy. P.275.

15. Creasy. P.282.

16. Creasy. P.290.

Chapter 9 Steam, Steel, and Modern War

1. Fuller. *Armament and History*. PP.106 – 109.

2. Fuller, *Armament and History* P.96.

3. Palit, P.36.

4. Fuller. *Armament and History* P.107.

5. Fuller. *Armament and History* P.111.

6. Fuller. *Armament and History* P.112.

7. Fuller. *Armament and History*. PP.111 – 113.

8. Fuller, . *Armament and History*. P.113.

9. Fuller. *Armament and History*. P.114.

10. Fuller. *Armament and History*. PP.9 – 10.

11. Fuller. *Armament and History*. P.11.

12. Fuller. *Armament and History*. P.11.

13. Cleator. P.176.

14. Cleator. P.177.

15. Palit. PP.87 – 88.

16. Fuller. *Armament and History*. P.140; palest, P.91.

17. Palit. P.89.

18. Palit. P.98.

19. Cleator. P.169.

20. Cleator. PP.170 – 171.

21. Cleator. P.171.

22. Cleator. P.170.

23. Cleator. P.194.

24, Cleator. PP.195 – 196.

Chapter 10 Challenge Number One – Overcoming the Power of Prevailing Fashion

1. Brennan, D. G. "Weaponry". *Toward the Year 2018*. (New York: Cowles education Corporation, 1968), PP.1 – 28.

2. Martino, Joseph. *Technological Forecasting for Decision Making* (New York: American Elsevier publishing company). P.82.

3. Holley, I. B. Jr. *Air Leadership: Proceedings of a Conference at Bolling Air Force Base April 13-14, 1984* . U.S. Air Force Academy. February, 2002. n.p. <https://books.google.com/books?id=LhtvCwAAQBAJ&pg=PT58&dq=Chenault+and+drop+tanks+for+fighter+aircraft&hl=en&sa=X&ved=0ahUKEwjZy9be2OTLAhWkrYMKHR67Dl4Q6AEINTAA#v=onepage&q=Chenault%20and%20drop%20tanks%20for%20fighter%20aircraft&f=false>

4. ibid

5. ibid

6. ibid

7. Beach, Lee Roy. *Making the Right Decision: Organizational Culture, Vision, and Planning.* Prentice Hall, 1993. Page 12

8. Beach, pp. 68-69

9. Hill, Charles W.L. and Gareth R. Jones, *Essentials of Strategic Management,* Houghton Mifflin Co. 2008. page 11

10. "Brilliant Pebbles". *Missile Threat: A Project of the George C. Marshall and Clarmont Institutes.* n.d. n. p. <http://missilethreat.com/defense-systems/brilliant-pebbles/>

11. ibid

12. ibid

13. ibid

Chapter 11 Challenge Number Two – Looking Accurately into the Future

1. Leavitt, Lloyd . "Initiatives in Target Acquisition and Destruction," *Readings for Reserve Forces Course.* Air Command and Staff College, 1978, PP.423 - 424.

Reference Notes

2. Possony and Pournelle, PP. 2-4.

3. Possony and Pournelle, PP.3 - 4.

4. Possony and Pournelle, PP.37.

5. Possony and Pournelle, PP.39, 40.

6. Possony and Pournelle, PP.41, 42.

7. Possony and Pournelle, PP.38, 42.

8, Brennan, P.4.

9. Possony and Pournelle, P.118.

10. Perry, Robert L. *The Ballistic Missile Decisions* (Santa Monica: the RAND Corporation, 1967). P.6.

11. Martino. P.71.

12. Valenti, Phillip "Leibniz, Pepin, and The Steam Engine: A Case Study of British Sabotage," *Fusion - Magazine of the Fusion Energy Foundation* December, 1979. PP.26 - 45.

13. Possony and Pournelle, PP.36 - 37.

14. Possony and Pournelle. P.103.

15. Martino. PP.103 - 122.

16. Martino. P.570.

17. Martino. PP.128 - 129.

18. Martino. P. 212.

19. Martino. P.212.

20. Martino. PP.214 - 215.

21. Martino. P.215.

22. Martino. PP.125 – 246.

Notes for Chapter 12 Today's Reality and Today's Fashion

1. "Missile defense systems by country", Wikipedia. n.d. n.pag.
<https://en.wikipedia.org/widi/Missile_de.ense_systems_by-country>

2. "Current counter-ICBM systems", *Anti-ballistic Missile* . n.d. n.pag.
<https://en.wikipedia.org/wiki/Anti-ballistic_missile>

3. "LGM-118 Peacekeeper". Wikipedia, n.d. n.pap.
<https://www.google.com/search?site=&source=hp&q=When+was+the+Peacekeeper+missile+first+operational&oq=When+was+the+Peacekeeper+missile+first+operational&gs_l=hp.12...4060.15852.0.21846.51.33.0.0.0.0.958.4718.5-2j4.6.0..2..0...1.1.64.hp..45.3.2407.0.mWeomgigGEo)>

4. ibid

5. "UGM-73 Poseidon". *Wikipedia*, n.d. n. pag. <https://en.wikipedia.org/wiki/UGM-73_Poseidon>

6. *Wikipedia*, n.d. n.pag. https://en.wikipedia.org/wiki/Submarine-launched_ballistic_missile)

7. "Borea-Class Submarine". n.d. n.pag. < https://en.wikipedia.org/wiki/Borei-class_submarine>

8. Scrowcroft, Lt Ben Brent. "Strategic Systems Development and New Technology: Where should we be going?" *New Technology and Western Security Policy,* Robert O'Neill, ed. International Institute for Strategic Studies, 1985 (978-0-333-39717-6) pp.5-6 <http://www.palgrave.com/us/book/9781349081943)>

9. ibid

10. Schneider, Mark . *The Nuclear Forces and Doctrine of the Russian Federation,* Foreword By Congressman Curt Weldon, Vice Chairman, House Armed Services Committee*, A Publication of the United States Nuclear Strategy Forum*, 2006 Washington, DC Publication No. 0003 © National Institute Press, 2006 *n.pag.* ISBN 0-

Reference Notes

9776221-0-X pp.

11. ibid

12. "New Start". *Wikipedia*. n.d. n.pag. <https://en.wikipedia.org/wiki/New_START>

13. ibid

14. ibid

15. Panda, Ankit . "US Prompt Global Strike Missiles Prompt Russian Rail-Mounted ICBMs: Russia will move ahead with an updated rail-mounted missile system for Inter-Continental Ballistic Missiles". *The Diplomat*, December 19, 2013. N.pag.<http://thediplomat.com/2013/12/us-prompt-global-strike-missiles-prompt-russian-rail-mounted-icbms/>

16. Ford, Christopher . "Does 'New START' Fumble Reloads and Rail-Mobile ICBMS?". *Hudson Institute* April 26th, 2010. n.pag. <http://www.hudson.org/research/9117-does-new-start-fumble-reloads-and-rail-mobile-icbms->

17. Numerical data is taken from *Soviet Military Power - 1985*, Superintendent of Documents, U.S. Government Printing Office, pages 27-38 with supplementation from *Wikipedia*. Brief quotations are from Soviet Military Power

18. "LGM-Minuteman III". *WMD Around the World*, Federation of American Scientists, n.d. n.pag. <http://fas.org/nuke/guide/usa/icbm/lgm-30_3.htm>

19. *Wikipedia,* n.d. n.pag. <https://en.wikipedia.org/wiki/Submarine-launched_ballistic_missile>

20. Thompson, Loren . "Some Disturbing Facts About America's Dwindling Bomber Force". *Forbes* Aug 16, 2013 n.pag.

<http://www.forbes.com/sites/lorenthompson/2013/08/16/some-disturbing-facts-about-americas-dwindling-bomber-force/#3d96ed7d757c>

21. "Russia", *National Threat Initiative (NTI)* . March, 2016 Nuclear n.pag. <http://www.nti.org/learn/countries/russia/nuclear/ > "Typhoon-class submarine", n.d. n.pag. < https://en.wikipedia.org/wiki/Typhoon-class_submarine > and, "Delta-class submarine" *Wikipedia.* n.d. n.pag. <https://en.wikipedia.org/wiki/Typhoon-

class_submarine>

22. ibid

23. Gady, Franz-Stefan, "How Many Ballistic Missile Submarines Will Russia Build?". *The Diplomat*. July 06, 2015. n.pag. <http://thediplomat.com/2015/07/how-many-ballistic-missile-submarines-will-russia-build/>

24. "Russia", *National Threat Initiative (NTI)*. March, 2016. Nuclear n.pag. <http://www.nti.org/learn/countries/russia/nuclear/ > "Typhoon-class submarine". *Wikipedia,* n.d. n.pag. <https://en.wikipedia.org/wiki/Typhoon-> class_submarine> -- and, "Delta-class submarine". *Wikipedia*. n.d. n.pag. <https://en.wikipedia.org/wiki/Typhoon-class_submarine>

25. *Soviet Military Power*, 1983, Superintendent of Documents, U.S. Government Printing Office,

26. ibid

27. Kristensen, Hans M. and Robert S. Norris "Russian Nuclear Forces, 2015", *Nuclear Notebook*, Federation of American Scientists, 1 May 2015 . n.pag. <http://thebulletin.org/2015/may/russian-nuclear-forces-20158299 >-

28. ibid

29. Kristensen, Hans M. and Robert S. Norris "Status of World Nuclear Forces", Federation of American Scientists, n.d. n.pag. <http://fas.org/issues/nuclear-weapons/status-world-nuclear-forces/>

30. Gertz, Bill . "Russia Doubling Nuclear Warheads", *The Washington Free Beacon*. April 1, 2016 n.pag. <http://freebeacon.com/national-security/russia-doubling-nuclear-warheads/>

31. ibid

32. ibid

33. ibid

34. ibid

35. ibid

36. ibid

37. *International Campaign to Abolish Nuclear Weapons (ican)* . n.d. n.pag.<http://www.icanw.org/the-facts/nuclear-arsenals/>

38. ibid

39. Peck, Michael . "Get Ready, America: Here Comes China's Ballistic Missile Defenses" *National Interest,* October 25, 2015. n.pag. <hhttp://nationalinterest.org/feature/get-ready-america-here-comes-chinas-ballistic-missile-14162>

40. MacDonald, Bruce W. and Charles D. Ferguson, "Understanding the Dragon Shield: Implications of Chinese Strategic Ballistic Missile Defense", A special report published by the *Federation of American Scientists*, September 2015. n.pag. <https://fas.org/pub-reports/understanding-the-dragon-shield/>

41. Rogoway, Tyler . "Why The USAF's Massive $10 Billion Global Hawk UAV Is Worth The Money", *Foxtrot Alpha*, September 9, 2014. n.pag.

<http://foxtrotalpha.jalopnik.com/why-the-usafs-massive-10-billion-global-hawk-uav-was-w-1629932000)>

42. "Russia tests Nudol anti-satellite system". *Russian strategic nuclear forces* April 1, 2016 n.pag. <http://russianforces.org/blog/2016/04/russia_tests_nudol_anti-satell.shtml> Gertz, Bill. "Russia Flight Tests Anti-Satellite Missile, Moscow joins China in space warfare buildup". *The Free Beacon*. December 2, 2015. n.pag. <http://freebeacon.com/national-security/russia-conducts-successful-flight-test-of-anti-satellite-missile/>

43. ibid

44. Gertz, Bill "Russia Flight Tests Anti-Satellite Missile" n.pag.

Notes for Chapter 13 The Impact of the Nuclear Bomb

"The atomic Bombs Saved 35 Million Lives", *Malaysian Neocon,* n.d. n.pag.
<https://scotthong.wordpress.com/2008/05/27/the-atomic-bombs-saved-35-million-lives/ >

2. ibid

3. ibid

4. Trueman, C. N. "Operation Downfall"
historylearningsite.co.uk. The History Learning Site, 19 May 2015. 3 Mar 2016. n.pag.
<http://www.historylearningsite.co.uk/world-war-two/the-pacific-war-1941-to-1945/operation-downfall/>

5. "How the Atomic bomb Saved 4,000,000 Lives" *Omaha World Herald*, November, 1987 Posted on 9/25/2006 n.pag.
<http://www.freerepublic.com/focus/news/1708051/posts>

6. ibid

7. ibid

8. ibid

9. ibid

10. "The Atomic Bombs Saved 35 Million Lives"

11. Compton, Karl T. "If the Atomic Bomb Had Not Been Used", The Atlantic, December 1946 Issue. n.pag.
<http://www.theatlantic.com/magazine/archive/1946/12/if-the-atomic-bomb-had-not-been-used/376238/>

12. "The Atomic Bombs Saved 35 Million Lives"

13. ibid

14. Miller, Henry I. "The Nuking Of Japan Was A Tactical And Moral Imperative".

Forbes, Aug 1, 2012. n.pag.

<http://www.forbes.com/sites/henrymiller/2012/08/01/the-nuking-of-japan-was-a-tactical-and-moral-imperative/#1b8450624881>

15. "How the Atomic bomb Saved 4,000,000 Lives"

16. Compton, Karl T. "If the Atomic Bomb Had Not Been Used

Notes for Appendix
The War Against Terror

1. "War on Terror", Wikipedia, n.d. n. pag.
<https://en.wikipedia.org/wiki/War_on_Terror>

2. ibid

3. Csaplar, Richard C. Jr. "1,400 Years of Christian/Islamic Struggle: An Analysis" CBN n.d. n.pag. <http://www1.cbn.com/churchandministry/1400-years-of-christian-islamic-struggle>

4. "First Crusade (1096-1099)" n.p. n.d. n.pag.

<http://www.umich.edu/~eng415/timeline/summaries/First_Crusade.htm>

5. Csaplar

6, ibid

7. Arlandson, James. "The Truth about Islamic Crusades and Imperialism". American Thinker . November 27, 2005 n.pag.
<http://www.americanthinker.com/articles/2005/11/the_truth_about_islamic_crusad.html >

8. Moore, John. "the evolution of Islamic terrorism: an overview", Target America, Frontline . n.d. n.pag. <http://www.pbs.org/wgbh/pages/frontline/shows/target/>

9. ibid

10. Mora, Edwin. "FBI Director Goes Where Obama Doesn't: Yes, It Is 'Radical Islamic' Terror", *Breitbart.com* December 10, 2015, n.pag. <http://www.breitbart.com/big-government/2015/12/10/fbi-director-goes-obama-doesnt-yes-radical-islamic-terror/>

11. DePetris, Daniel R. "The 5 Deadliest Terrorist Groups on the Planet", *The National*

Interest, November 16, 2014, n.pag. <http://nationalinterest.org/feature/washington-watching-the-5-deadliest-terrorist-groups-the-11687>

12. Pagliery, Jose . "ISIS is struggling to fund its war machine", *CNN Money*, April 21, 2016 n.pag. <http://money.cnn.com/2016/04/21/news/isis-financing-oil-tax-war/>

13. DePetris

14. Ferran, Lee and Rym Momtaz. "ISIS: Trail of Terror" *ABC News*. n.d. n.pag. <http://abcnews.go.com/WN/fullpage/isis-trail-terror-isis-threat-us-25053190>

15. ibid

16. Hoffman, Bruce. "The Coming ISIS - al Qaeda Merger: It's Time to Take the Threat Seriously", *Foreign Affairs*, March 29, 2016. n.pag. <https://www.foreignaffairs.con/articles/2016-03-29/coming-isis-al-qaeda-merger>

17. Weiss, Martin. "Nuclear terror predicted by 3 major studies". Money and Markets, April 11, 2016. PP. 1-3.

18. Ferran, Lee, and Rym Momtaz. "ISIS: Trail of Terror" *ABC News* n.d. n.pag. <http://abcnews.go.com/WN/fullpage/isis-trail-terror-isis-threat-us-25053190>

19. Kuperman, Alan J. "Obama's Libya Debacle: How a Well-Meaning Intervention Ended in Failure". Foreign Affairs . March/April 2015 n.pag. <https://www.foreignaffairs.com/articles/libya/obamas-libya-debacle>

20. ibid

21. Baroudos, Constance "Chaos in Libya: The Rising ISIS Threat to Europe" *The National Interest* . April 14, 2016 n.pag. <http://nationalinterest.org/blog/chaos-libya-the-rising-isis-threaten-europe-15801>

22. Bunn, Matthew, Martin B. Malin, Nickolas Roth, and William H Tobey, "Advancing Nuclear Security: Evaluating Progress and Setting New Goals," *Harvard Kennedy School, Belfer Cener for Science and International Affairs,* March 2014.

23, Yourish, Karen, Tim Wallace, Derek Watkins, Tom Biratikanon, "Brussels is Latest Target in Islamic State's Assault on West" *New York Times,* March 25, 2016. N.pag. <http://www.nytimes.con/interactive/2016/03/25/world/map-isis-attacks-around-the-world.html > (List of Major ISIS attacks) <https://www.google.com/search?site=&source=hp&q=brussels+is+latest+target+in+islamic+state%E2%80%99s+assault+on+west&oq=Brussels+is+latest+target+in+islamic+state&gs_l=hp.1.0.0.4443.15151.0.17509.43.39.0.4.4.0.178.4876.0j38.38.0.…0…1c.1.64.hp..1.40.4653.0..35i39j0i131j0i155i3j0i20j0i22i30.kANDY_tkw0g>

24. Nossiter , Adam. "Abuses by Nigeria's Military Found to Be Rampant in War Against Boko Haram" *The New York Times*. June 3, 2015. n.pag. http://www.nytimes.com/2015/06/04/world/africa/abuses-nigeria-military-boko-haram-war-report.html?_r=0 ;

25. Cooper, Helene. "Boko Haram and ISIS Are Collaborating More, U.S. Military Says" *New York Times* . April 20, 2016. n.pag. <http://www.nytimes.com/2016/04/21/world/africa/boko-haram-and-isis-are-collaborating-more-us-military-says.html?_r=0>

26, Nossiter

27. Nossiter

28, Nossiter

29. Masters, Jonathan, and Zachary Laub, "CFR Backgrounders". Council on Foreign Relations. January 3, 2014. n.pag. http://www.cfr.org/lebanon/hezbollah-k-hizbollah-hizbullah/p9155

30. "Muslim Brotherhood Wikipedia", Wikipedia. n.d. n.pag. <https://en.wikipedia.org/wiki/Muslim_Brotherhood>

31. Laub, Zachary . "Egypt's Muslim Brotherhood"
CFR Backgrounders. Council on Foreign Relations. January 15, 2014. n.pag. <http://www.cfr.org/egypt/egypts-muslim-brotherhood/p23991>

32. Beamon, Todd, "UK Declares Muslim Brotherhood Terrorist Group, Breaks With Obama". Newsmax. 22 Dec 2015. n.pag. <http://www.newsmax.com/Newsfront/uk-

declare-muslim-brotherhood-terrorist/2015/12/22/id/706848/>

33. Ibid

34. Laub, Zachary. "Hamas" CFR Backgrounders, Council on Foreign Relations August 1, 2014. n.pag.
<http://www.cfr.org/israel/hamas/p8968> 35. Hayes, Laura , Borgna Brunner, and Beth Rowen. "Osama bin Laden's Network of Terror Al-Qaeda" n.p. n.d. n.pag.
http://www.infoplease.com/spot/al-qaeda-terrorism.html>

36. ibid

37. "Country Reports on Terrorism 2014" *Department of State*, April 2015, n.pag. <http://www.state.gov/j/ct/rls/crt/2014/>

38. Issacharoff, Avi. "Boosted by nuke deal, Iran ups funding to Hezbollah, Hamas". *The Times of Israel*. September 21, 2015. n.pag.
<http://www.timesofisrael.com/boosted-by-nuke-deal-iran-ups-funding-to-hezbollah-hamas/>

39. Bergen, Peter. "How big is the U.S. terror threat?". CNN . March 23, 2016. n.pag. <http://www.cnn.com/2016/03/22/opinions/terrorism-threat-united-states-bergen/>

40. ibid

41. Bunn, Matthew, Martin B. Malin, Nickolas Roth, and William H Tobey, "Advancing Nuclear Security: Evaluating Progress and Setting New Goals". Harvard Kennedy School, Belfer Cener for Science and International Affairs. March 2014, n.pag.

42. ibid

43. ibid

44. 'FBI Investigating Terrorism in All 50 States" ," *The Clarion Project*. July 23, 2015. n.pag. <http://www.clarionproject.org/news/fbi-investigating-terroris-all-50-states?gclid-CjOKEQjwl-e4BRCwqeWkv8TWqOoBEiQAMocbP6MPu-UG1UeKy-Zml6Cgktn6wJSZZRuOpFmMsrJWIKAaAmMS8PHAQ#>

45. Kaplan, Eben. "Terrorists and the Internet" *Council on Foreign Relations*, January

8, 2009. n.pag. < http://www.cfr.org/terrorism-and-technology/terrorists-internet/p10005>

46. Schorn, Daniel. "Terrorists Take Recruitment Efforts Online" (Report of 'Sixty Minutes' broadcast with Scott Pelley) *CBS News*. March 2, 2007, n.pag. <http://www.cbsnews.com/news/terrorists-take-recruitment-efforts-online/>

47. "The Internet as a Terrorist Tool for Recruitment and Radicalization of Youth" White Paper. HIS Publication Number: RP08-03.02.17-01, U.S. *Department of Homeland Security*, Science and Technology Directorate. 24 April, 2009. n.pag. <https://www.google.com/search?site=&source=hp&q=the+internet+as+a+terrorist+tool+for+recruitment+and+radicalization+of+youth&oq=The+internet+as+a+terrorist+tool&gs_l=hp.1.0.0j0i22i30.4477.13509.0.16809.33.31.0.2.2.0.224.4246.0j28j2.30.0....0...1c.1.64.hp..1.30.4011.0..35i39j0i131j0i20j0i3j0i22i10i30.jsw7iHF8I2Q>

48. Howell, Tom Jr. "Homeland Security chief Jeh Johnson says no credible terrorist threat to U.S. ". *The Washington Times*, November 22, 2015, n.pag. <http://www.washingtontimes.com/news/2015/nov/22/jeh-johnson-homeland-security-chief-says-no- credib/>

49, Pop, Valentina. "Belgium Says Hundreds Fighting in Syria Pose Europe Threat". *The Wall Street Journal*. April 21, 2016. n.pag. <wsj.com/articles/Belgium-says-200-belgians-fighting-in -syria-pose-terror-threat-1461237039>

50. Bergen, Peter. "How big is the U.S. terror threat?" *CNN*. March 23, 2016. n.pag. < http://www.cnn.com/2016/03/22/opinions/terrorism-threat-united-states-bergen/>

51. de Winter, Leon. "Europe's Muslims hate the West". *Politico Europe Edition* March 29, 2016. n.pag. <http://www.politico.eu/article/brussels-attacks-terrorism-europe-muslims-brussels-attacks-airport-metro/>

52. Peters, Ralph. "Our New Old Enemies". From *Parameters,* Summer 1999, pp. 22-37.
https://www.google.com/search?site=&source=hp&q=ralph+peters+our+new+old+enemies&oq=Ralph&gs_l=hp.1.0.35i39j0j0i20l2j0l6.3395.4374.0.6530.6.6.0.0.0.0.167.583.0j4.4.0....0...1c.1.64.hp..2.3.413.0.RTKGz_P0jl4

The author has three decades of experience as a Development Engineer, Operations Research Analyst, and staff officer in the Research and Development field, working on advanced technology concepts ranging from aerospace defense and airborne high energy lasers to tactical aircraft. He then was a university professor. He earned an engineering degree at Texas Tech and an MBA and the Ph.D. at The Ohio State University. He has authored a dozen papers for national symposia, another dozen journal articles, and two previous books.

Made in the USA
Lexington, KY
19 January 2017